A First Course
in
Structural Equation Modeling

A First Course
in
Structural Equation Modeling

Tenko Raykov
Fordham University

and

George A. Marcoulides
California State University, Fullerton

LAWRENCE ERLBAUM ASSOCIATES, PUBLISHERS
2000   Mahwah, New Jersey   London

Lawrence Erlbaum Associates, Inc., Publishers
10 Industrial Avenue
Mahwah, NJ 07430

Cover design by Kathryn Houghtaling Lacey

Library of Congress Cataloging-in-Publication Data

Raykov, Tenko.
A first course in structural equation modeling / Tenko Raykov and George Marcoulides.
   p. cm.
Includes bibliographical references and index.
ISBN 0-8058-3568-7 (cloth : alk. paper) — ISBN 0-8058-3569-5 (pbk. : alk. paper)
   1. Multivariate analysis. 2. Social sciences—Statistical methods. I. Marcoulides, George, A.   II. Title.
QA278.R39   2000
519.5'35—dc21   99-056749
                                                    CIP

Printed in the United States of America
10 9 8 7 6 5 4 3 2 1

Contents

Preface

We wrote this book for an introductory structural equation modeling (SEM) course similar to the ones we teach at Fordham University and California State University, Fullerton. Our goal is to present a conceptual and nonmathematical introduction to SEM methodology. The readership we have in mind consists mainly of graduate students or researchers from any discipline with limited or no previous exposure to SEM. When we examined other available books, we found that most of them had serious limitations that precluded their use in an introductory course. These books were either too technical for beginners, did not cover in sufficient breadth the basics of the methodology that we consider to be relevant for this type of course, or intermixed fairly advanced issues with basic ones. Our book is therefore an alternative attempt to provide a first course in SEM methodology at a coherent introductory level.

There are no special prerequisites for readers of this text beyond a course in basic statistics that included coverage of regression analysis. Because we frequently draw a parallel between aspects of SEM and their apparent analogs in regression, this prior knowledge is important. There are also only a few mathematical formulas used, which are either conceptual or illustrative rather than computational in nature. Although the basic ideas and methods for conducting SEM analyses presented in this book are independent of the particular computer programs used, for clarity the examples are illustrated using the two apparently most widely circulated programs, LISREL and EQS. As such, the book can also be used as a first guide for learning how to set up input files to fit several frequently used types of structural equation models with these programs.

Due to the targeted audience of first-time SEM users, many important advanced topics could not be included in the book. Anyone interested in such topics can consult a plethora of more advanced SEM texts published throughout the last decade (*http://www.erlbaum.com*). Our book can be considered a helpful precursor to these advanced SEM texts.

Our efforts to produce this text would not have been successful without the continued support and encouragement provided by a number of scholars in the SEM area. We feel particularly indebted to Peter M. Bentler, Michael W. Browne, Karl G. Jöreskog, and Bengt O. Muthén for their important contributions to the methodology and their helpful discussions and instruction throughout the years. In many regards they have profoundly influenced our understanding of SEM and this is reflected in the book. We are grateful to Paul B. Baltes, Freyda Dittmann-Kohli, and Reinhold Kliegl for the permission to use data from their project on aging and plasticity in fluid intelligence of older adults for illustrative purposes. We would also like to thank numerous colleagues and students who provided comments on earlier drafts of the various chapters. Thanks are also due to all the wonderful people at Lawrence Erlbaum Associates for their assistance and support. Finally, and most importantly, we thank our families for their continued love and support despite the fact that we keep taking on new projects. Specifically, the first author wishes to thank Albena and Anna; and the second author wishes to thank Laura and Katerina.

—Tenko Raykov
—George A. Marcoulides

CHAPTER ONE

Fundamentals of Structural Equation Modeling

WHAT IS STRUCTURAL EQUATION MODELING?

Structural equation modeling is a statistical methodology used by biologists, economists, educational researchers, marketing researchers, medical researchers, and a variety of social and behavioral scientists. One reason for its pervasive use in many scientific fields of study is that structural equation modeling provides researchers with a comprehensive method for the quantification and testing of theories. Other major characteristics of structural equation models are that they explicitly take into account the measurement error that is ubiquitous in most disciplines and contain latent variables.

Latent variables are theoretical or hypothetical constructs of major importance in many sciences. Typically, there is no direct operational method for measuring latent variables or a precise method for assessing their degree of presence. Nevertheless, manifestations of a construct can be observed by recording or measuring specific features of the behavior on some set of subjects in a particular environment. The recording or measurement of features of the behavior is typically obtained by using pertinent instrumentation (e.g., tests, scales, self-reports, inventories, or questionnaires). Once constructs have been assessed, structural equation modeling can be used to test the plausibility of hypothetical assertions about potential interrelationships among the constructs as well as their relationships to the indicators or measures assessing them. Due to the mathematical complexities of estimating and testing the proposed assertions, computer programs are a must in applications of structural equation modeling methodology. To date, numerous computer programs are available for conducting structural equation modeling analyses. Programs such as AMOS (Arbuckle, 1995), EQS (Bentler, 1995), LISREL (Jöreskog & Sörbom, 1993a, 1993b, 1993c, 1999), MPLUS (Muthén & Muthén, 1998), SAS-PROC CALIS (SAS Institute, 1989), SEPATH (Statistica, 1998), and RAMONA (Browne & Mels, 1994) are likely to contribute to a further increase in the coming years of the popularity of this relatively new methodology. Although all these programs have somewhat similar capabilities, LISREL and EQS have historically dominated the field (Marsh, Balla, & Hau, 1996). For this reason (and because it would be impossible to cover every program in detail in an introductory text), all the examples in this book are illustrated using only the LISREL and EQS programs.

The term *structural equation modeling* (SEM) is used throughout this book as a generic term to refer to various types of commonly encountered models. The following are some characteristics of SEM models.

1. The models are usually conceived in terms of not directly measurable, and possibly not (very) well-defined, theoretical or hypothetical constructs. For example, anxiety, attitudes, goals, intelligence, motivation, personality, and socioeconomic status are all representative of such constructs.

2. The models usually take into account potential errors of measurement in all variables. This is achieved by including an explicit error term for each fallible measure, whether it is an explanatory or predicted variable. The variances of the error terms are, in general, parameters that must be estimated when any model is fit to data. Tests of hypotheses about them can also be conducted when they represent substantively meaningful assertions about the error terms or their relationships to other parameters.

3. The models are usually fit to matrices of interrelationship indices (i.e., covariance or correlation matrices) between all pairs of observed variables, and sometimes to variable means.[1]

---

[1]It can be shown that the fit function minimized in the maximum likelihood method, used in the majority of current applications of SEM, is based on the likelihood function of the raw data (see Bollen, 1989; see also the section Rules for Determining Model Parameters). Thus, with multinormality, a structural model can be considered indirectly fitted to the raw data as well, similar to models within the general linear modeling framework. Because this is an introductory book, however, we emphasize here the more direct process of fitting the model to the analyzed matrix of variable interrelationship indices, which is the underlying idea of the most general asymptotically distribution-free method of model fitting and testing. The maximization of the likelihood function of the raw data is equivalent alternatively to the minimization of the fit function of the maximum likelihood method, which measures the distance between that matrix and the one reproduced by the model (see the section Rules for Determining Model Parameters).

This list of characteristics can also be used to differentiate structural equation models from classical general linear modeling approaches. The classical approaches encompass such methods as regression analysis, analysis of variance, analysis of covariance, and a large part of multivariate statistical methods (see Marcoulides & Hershberger, 1997). In the classical approaches, models are fit to raw data, and no error of measurement in the independent variables is assumed.

On the other hand, an important feature that many of the classical approaches share with SEM is that they are based on linear models. Thus, a frequent assumption when using SEM methodology is that the relationships between observed variables are linear (although modeling nonlinear relationships is gaining popularity in SEM; see Schumacker & Marcoulides, 1998). Another shared property between classical approaches and SEM is model comparison. For example, the F test for comparing a less restricted model to a more restricted model is used in regression analysis whenever a special case of a given model is tested. An example of such a case might be when a researcher is interested in testing whether to drop from a prediction model one or more predictors. As discussed later, the counterpart of this test in SEM is the difference in chi-square values test, or its asymptotic equivalents in the form of Lagrange multiplier or Wald tests (Bentler, 1995). Generally speaking, the chi-square test is used in SEM to examine the plausibility of model parameter restrictions such as equality of factor loadings, factor or error variances, or factor variances and covariances across groups.

Types of Structural Equation Models

The following types of commonly used structural equation models are considered in this book.

1. *Path analysis models.* Path analysis models are usually conceived only in terms of observed variables. For this reason, some researchers do not consider path analysis models to be typical SEM models. We believe that path analysis models are worthy of discussion within the general SEM framework because, although they only focus on observed variables, they are an important part of the historical development of SEM and use the same underlying idea of model fitting and testing as any other SEM model. Figure 1 presents an example of a path analysis model examining the effects of some variables on the number of hours people spend watching television (see the section Path Diagrams for a complete list and discussion of the symbols that are commonly used to graphically represent SEM models).

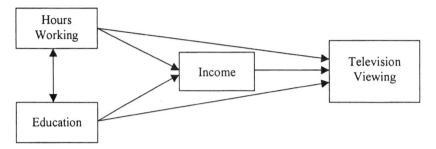

FIG. 1.   Path analysis model examining the effects of some variables on television viewing. Education = Number of completed school years; Hours Working = Average weekly working hours; Income = Yearly gross income in dollars; Television Viewing = Average daily number of hours spent watching television.

2. *Confirmatory factor analysis models.* Confirmatory factor analysis models are commonly used to examine patterns of interrelationships among several constructs. Each construct included in the model is usually measured by its own set of observed indicators. Thus, in a confirmatory factor analysis model no specific directional relationships are assumed between the constructs, only that they are correlated with one another. Figure 2 presents an example of a confirmatory factor analysis model with

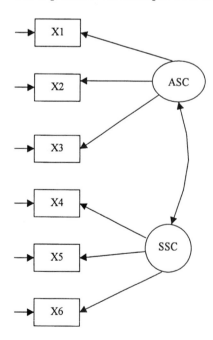

FIG. 2.   Confirmatory factor analysis model with two self-concept constructs. ASC = Academic self-concept; SSC = Social self-concept.

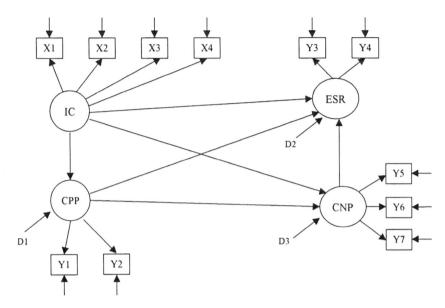

FIG. 3.   Structural regression model of variables influencing return to promotion. CNP = Characteristics of new positions; CPP = Characteristics of prior positions; ESR = Economic and social returns to promotion; IC = Individual characteristics.

two interrelated self-concept constructs (Marcoulides & Hershberger, 1997).

3. *Structural regression models.* Structural regression models resemble confirmatory factor analysis models, except that they also postulate specific explanatory relationships (latent regressions) among constructs. The models can be used to test or disconfirm proposed theories about explanatory relationships among various latent variables under investigation. Figure 3 presents an example of a structural regression model of variables influencing returns of promotion for faculty in higher education (Heck & Johnsrud, 1994).

4. *Latent change models.* Latent change models represent a way to study change over time. The models focus primarily on patterns of growth, decline, or both in longitudinal data (e.g., on aspects such as initial status and rates of change over time) and enable researchers to examine both intra- and interindividual differences in patterns of change. The models can also be used to examine the relationships between patterns of change and other personal characteristics. Figure 4 presents an example of a simple two-factor growth model for two time points.

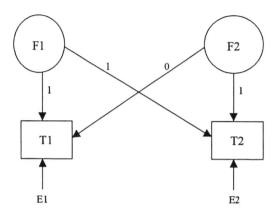

FIG. 4. A simple latent change model (error variances assumed equal).

When and How Are Structural Models Used?

Structural equation models can be used to represent knowledge about phenomena studied in particular substantive domains. The models are usually based on proposed theories that describe and explain the phenomena under investigation. With its unique feature of explicitly including measurement error, structural models provide an attractive means for achieving this goal. Once a theory has been developed about some phenomena of interest, it can be tested against empirical data. This process of testing is often called the *confirmatory mode* of SEM applications.

A related application of structural models is *construct validation*. In this application, a researcher is interested mainly in evaluating the extent to which a particular instrument actually measures the latent variable(s) it is supposed to measure. This type of application is most frequently used when studying the psychometric properties of a particular measurement device.

Structural models are also used for theory development. In *theory development*, the process often involves repeated applications of SEM on the same data set in order to explore potential relationships between latent variables of interest. In contrast to the confirmatory mode of SEM applications, theory development assumes that no prior theory exists—or that one exists only in a rudimentary form—about the phenomena of interest. Because this application of SEM contributes both to the clarification and development of theories, it is commonly referred to as the *exploratory mode* of SEM applications. It is important to note that because theory development is generally carried out on a single data set (from one sample), results from such exploratory applications of SEM need to be interpreted with caution. Only when the results are replicated across other samples from the same population of interest can they be considered trustworthy. The reason for this concern stems mainly from the fact that the results obtained by repeated SEM applications may be capitalizing on chance factors within the analyzed sample, thereby limiting the generalizability of the model beyond that sample.

Why Are Structural Equation Models Used?

A main reason that structural equation models are widely used in many scientific fields of study is that they provide a mechanism for explicitly taking into account measurement error in the observed variables (both dependent and independent) considered in a model. In contrast, traditional regression analysis effectively ignores potential measurement error in all the explanatory (independent) variables included in a model. As a result, regression estimates can be misleading and potentially lead to incorrect substantive conclusions.

In addition to handling measurement error, SEM models also enable researchers to study both the direct and indirect effects of the various variables included in a model. *Direct effects* are the effects that go directly from one variable to a second variable. *Indirect effects* are the effects between two variables that are mediated by one or more intervening variables (often referred to as a mediating variable). The combination of direct and indirect effects make up the *total effect* of the explanatory variable on a dependent variable. Thus, if an indirect effect does not receive proper attention, the relationship between two variables of interest may not be fully considered. Although regression analysis can also be used to estimate indirect effects (e.g., by regressing the mediating variable on the explanatory variable, and then regressing the effect variable on the mediating variable and multiplying the regression weights) the approach is appropriate only when there are no measurement errors in the predictor variables. Such an assumption, however, is unrealistic in practice. In addition, standard errors for the estimates are difficult to compute using the sequential application of regression analysis, but are quite straightforwardly obtained in SEM.

What Are the Key Elements of Structural Models?

The key elements of essentially all structural models used in research practice are their parameters (often referred to as model parameters or unknown parameters). Model parameters reflect those aspects of a model that are usually unknown to the researcher, at least at the beginning of the analysis, yet are necessary for testing the model. *Parameter* is a generic term referring to some characteristic of a whole population (such as a mean or a variance). Although the characteristic of the whole population is difficult to obtain, it is essential in order to understand the phenomenon under investigation. Sample statistics are a means of estimating the parameter(s). In structural equation modeling, the parameters are unknown aspects of the studied phenomenon that are related to the distribution of the variables considered in a model. They have to be estimated, most often from the sample covariance or correlation matrix using specialized computer programs.

The presence of parameters in structural equation models should not pose any new difficulties. Regression analysis models also contain parameters. For example, the standard error of estimate and the regression weights (slopes) are model parameters. Similarly, in a factorial analysis of variance the main effects and interaction effects are considered the model parameters. In general, parameters are essential elements for all statistical models applied in practice. The parameters reflect the unknown aspects of the underlying phenomenon in a study and are estimated by fitting the model to sampled data using specific optimality criteria, numeric routines, and specialized software. The topic of model parameters along with a complete description of the rules used in SEM are discussed extensively in the section Parameter Estimation.

PATH DIAGRAMS

One of the easiest ways to communicate a structural equation model is to draw a picture of it. Pictures of SEM models are called *path diagrams* and are drawn using special graphical notation. A path diagram is a sort of mathematical representation (but in graphical form) of a model under investigation. As it turns out, path diagrams not only enhance the understanding of structural models, but they substantially contribute to the creation of the correct input files necessary to fit and test each proposed model using a particular software package. Figure 5 presents the most commonly used graphical notation for the representation of SEM models (described in detail in this section).

Latent and Observed Variables

One of the most important issues in SEM is the distinction between observed variables and latent variables. *Observed variables* are the variables that are actually measured, such as manifested performance on a particular test or the answers to specific items or questions on an inventory or questionnaire. The term manifest variables is also often used for these to stress the fact that these are the variables that have actually been measured

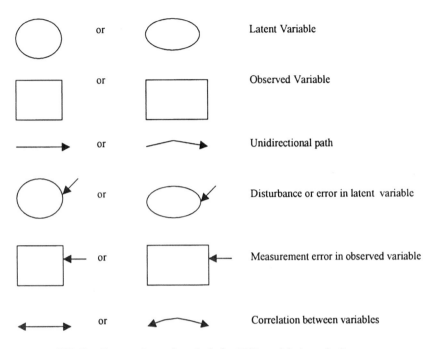

FIG. 5.   Commonly used symbols for SEM models in path diagrams.

by the researcher in the process of data collection. In contrast, *latent variables* are the hypothetically existing constructs of interest in a study. For example, intelligence, anxiety, locus of control, organizational culture, motivation, depression, social support, and socioeconomic status are all latent variables. The main characteristics of latent variables are that they cannot be measured directly (because they are typically unobservable directly) and, hence, only proxies for them can be obtained using specifically developed measuring instruments—tests, inventories, questionnaires, and so on. These proxies are the indicators of the latent constructs or, in simple terms, their measured aspects. For example, socioeconomic status may be measured in terms of income level, years of education, bank savings, type of occupation, and so on. Obviously, it is quite common for manifest variables to be fallible and unreliable indicators of the unobservable latent constructs. If a single observed variable is used as an indicator of a latent variable, it is most likely that the observed variable will contain quite unreliable information about the construct. This information can be considered to be one-sided because it reflects only one aspect of the measured construct (the side captured by the manifest variable used for measurement). It is therefore generally recommended that researchers use multiple indicators (preferably more than two) for each latent variable considered in order to obtain a much more complete and reliable picture than that provided by a single indicator. Of course, there are instances in which a single observed variable can be a very good indicator of a latent variable; for example, if the construct of intelligence is included in the model and the total score on the Stanford–Binet Intelligence Test is used as the indicator.

Squares and Rectangles, Circles, and Ellipses

Observed and latent variables are represented in path diagrams by two distinct graphical symbols. Squares or rectangles are used for observed variables and circles or ellipses are used for latent variables. Observed variables are usually labeled sequentially (e.g., X_1, X_2, X_3) with the label centered in each square or rectangle. Latent variables can be abbreviated according to the construct represented (e.g., *SES* for socioeconomic status) or just labeled sequentially (e.g., F_1, F_2; F for factor) with the name or label centered in each circle or ellipse.

Paths and Two-Way Arrows

Latent and observed variables are connected in a structural equation model in order to reflect a set of theoretical propositions about a studied phenomenon. Typically, the interrelationships among both observed and latent variables are the main focus of study. These relationships are represented graphically in a path diagram by one-way and two-way arrows. The one-way arrows, also called paths, signal that a variable at the end of the arrow is explained in the model by the variable at the beginning of the arrow. One-way arrows are usually represented by straight lines, with arrowheads at the end of the straight lines. It is important to note that such paths are often interpreted as symbolizing causal relationships—the variable at the end of the arrow is assumed to be the effect and the one at the beginning is assumed to be the cause. We believe that such inferences should not be made from path diagrams without a strong rationale for doing so. For instance, latent variables are frequently considered to be causes for their indicators; that is, the measured or recorded performance is considered to be the effect of the presence of the corresponding latent variable. Apart from these situations, however, we abstain from making causal interpretations except possibly when the variable considered temporally precedes another one (in which case the former could be interpreted as the cause of the one occurring later; see Babbie, 1992, chap. 1; Bollen, 1989, chap. 3). Speculations about the nature of causality in SEM models abound in the literature. Bollen (1989) lists three conditions that should be used to establish a causal relation between variables—isolation, association, and direction of causality. While association and direction of causality may be fairly easy to examine, it is quite difficult to ensure that a cause and effect have been isolated from all other influences. For this reason, many researchers consider SEM models and the causal relations within the model as approximations to reality that can never really be proved. They can only be disproved or disconfirmed.

Two-way arrows (sometimes referred to as two-way paths) are used to represent covariation between two variables and signal that there is an association between the connected variables that is not assumed to be directional. Usually two-way arrows are graphically represented as curved lines with an arrowhead at each end. A straight line with arrowheads at each end is also sometimes used to represent a correlation between variables (usually because of a lack of space). Lack of space may also force researchers to even represent a one-way arrow by a curved rather than a straight line, with an arrowhead attached to the appropriate end (see Fig. 5). Therefore, when first looking at a path diagram it is essential to determine which of the straight or curved lines have two arrowheads and which only one.

Dependent and Independent Variables

In order to accurately convey information about the structure of a proposed model to SEM programs like LISREL or EQS, there is another distinction between variables in a model that is of great importance—the differentiation between dependent and independent variables. *Dependent variables* are variables that receive at least one path (one-way arrow) from another variable in the model. *Independent variables* are variables that emanate paths, but never receive them. Independent variables can be correlated among one another (i.e., connected in the path diagram by two-way arrows). It is important to note that a dependent variable may act as an independent variable with respect to one variable, but this does not change its dependent-variable status. As long as there is at least one path ending at the variable, it is considered to be a dependent variable, no matter how many other dependent variables in the model are explained by that variable.

In the econometric literature, the terms exogenous variables (for independent variables) and endogenous variables (for dependent variables) are also frequently used to make the same distinction between variables. (These terms are derived from the Greek words *exo* and *endos*, for being "of external origin" to the system of variables under consideration and "of internal origin" to it.) Regardless of the terms one uses, an important implication of the distinction between dependent and independent variables is that there are no two-way arrows connecting any two dependent variables in a model path diagram. For reasons that will become much clearer later, the variances and co-variances (correlations) between dependent variables are explained within the model in terms of the unknown parameters.

An Example Path Diagram

To clarify further the discussion of path diagrams, consider the factor analysis model displayed in Fig. 6, representing the relationships among Parental dominance, Child intelligence, and Achievement motivation.

As can be seen by examining Fig. 6, there are nine observed variables in this model. The observed variables represent nine scale scores that were obtained from a sample of 245 elementary school students. The variables are denoted by the labels V_1 through V_9 (using V for observed Variable). The latent variables (or factors) are Parental dominance, Child intelligence, and Achievement motivation. As latent variables (or factors), they are denoted as F_1, F_2, and F_3, respectively. The latent variables are each

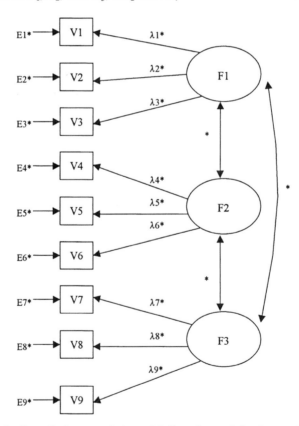

FIG. 6.   Example factor analysis model. F_1 = Parental dominance; F_2 = Child intelligence; F_3 = Achievement motivation.

measured by three indicators with each path representing the factor loading of the observed variable on the latent variable.

The "curved" two-way arrows in Fig. 6 symbolize the correlations between the latent variables (i.e., the factor correlations) in the model.
There is also a residual term attached to each manifest variable. The residuals are denoted by E (for Error), followed by the index of the variable to
which they are attached. Each *residual* represents the amount of variation
in the manifest variable that is due to measurement error or that remains
unexplained by variation in the corresponding latent factor the variable
loads on. The unexplained variance is the amount of indicator variance
unshared with the other measures of the particular common factor.

As indicated previously, it is important to determine which are the dependent and which are the independent variables of the model. As can be
seen in Fig. 6 (and using the definition of error), there are a total of 12 independent variables. These are the three latent variables and nine error
terms. It is important to note that there are no direct paths going to these
independent variables, but there are paths leaving each of them. In addition, there are three two-way arrows (representing the three correlations)
connecting the latent variables. The dependent variables are the nine observed variables labeled V_1 through V_9. These dependent variables receive
two paths going to them—the path from the latent variable it loads on
(representing its factor loading) and the one from its residual term (representing the error term).

First write the *model definition equations*. These are the relationships
between observed and unobserved variables that formally define the proposed model. Following Fig. 6, the equation for each of the dependent
variables in the model is obtained by writing an equation for each observed variable in terms of how it is explained in the model (i.e., in terms
of its latent variables and its corresponding error term). Based on this approach, the following system of nine equations is obtained (one for each
dependent variable in the model):

$$
\begin{aligned}
V_1 &= \lambda_1 F_1 + E_1, \\
V_2 &= \lambda_2 F_1 + E_2, \\
V_3 &= \lambda_3 F_1 + E_3, \\
V_4 &= \lambda_4 F_2 + E_4, \\
V_5 &= \lambda_5 F_2 + E_5, \\
V_6 &= \lambda_6 F_2 + E_6, \\
V_7 &= \lambda_7 F_3 + E_7, \\
V_8 &= \lambda_8 F_3 + E_8, \\
V_9 &= \lambda_9 F_3 + E_9,
\end{aligned}
\tag{1}
$$

where λ_1 to λ_9 denote the factor loadings that will be estimated based on
the observed data.

According to the factor analysis model under consideration, each of the nine Equations (1) represents the corresponding observed variable as the sum of the product of that variable's factor loading with its pertinent factor and a residual term. Note that on the left-hand side of each equation there is only one variable, the dependent variable, rather than a combination of variables, and that no independent variable appears there. It is also important to note that with this model the following assumptions are made: (a) that in the studied population the mean of each residual variable vanishes (i.e., is 0), and (b) that the residual terms are independent of the latent variables (i.e., errors are uncorrelated with the factors). Although these assumptions apply to any structural equation model, they are not unique to SEM. For example, they are also routinely made when conducting a regression analysis—the error in the regression equation is assumed to be uncorrelated with any predictor and has a mean of 0 in the population (Pedhazur & Smelkin, 1991).

Asterisks and Model Parameters

Another important feature of path diagrams, are the asterisks associated with one-way and two-way arrows (see Fig. 6). These asterisks are symbols of the unknown parameters and are very useful for properly running the model-fitting and estimation process of most SEM programs. It is therefore very important for researchers to be able to convey which are the unknown model parameters. If this is done incorrectly or arbitrarily, there is a danger of ending up with a model that is too restrictive or that has parameters that cannot be uniquely estimated. Such problematic parameter estimation is characteristic of models that are called unidentified (discussed in greater detail in a later section).

The requirement of explicitly stating the model parameters is quite unique to a SEM analysis. For example, in regression analysis with popular software one does not need to explicitly present the parameters of a fitted model. Suppose a researcher were interested in fitting the following regression model predicting levels of depression among college students:

$$\text{Depression} = a + b_1\text{Social\_Support} + b_2\text{Intelligence} + b_3\text{Age} + \text{Error}, \tag{2}$$

where a is the intercept and b_1, b_2, and b_3 are the partial regression weights (slopes). When this model is submitted to a major statistical package for analysis (e.g., SAS or SPSS), the researcher is not required to specifically define a, b_1, b_2, and b_3 as the model parameters. This is due to the fact that, unlike SEM, a regression analysis can be conducted in only one way with regard to the set of parameters. When a regression analysis is carried out, a researcher only needs to provide the program with information about which variables are to be used as explanatory variables and which as the dependent variables. The program then automatically determines the model parameters, typically one slope per predictor (partial regression weight) plus an intercept for the fitted regression equation.

This automatic or default determination of model parameters does not work well in SEM applications. It is particularly important in SEM to explicitly state the parameters to understand and correctly set up the model one is interested in fitting. Therefore, we strongly recommend that users always first define the parameters of each model they consider. Using default settings in SEM programs will not absolve a researcher from having to think carefully about the details of the particular model being examined. It is the researcher who must decide exactly how the model is defined, not the default features of a computer program. For example, if a factor analytic model similar to the one presented in Fig. 6 is being considered in a study, one researcher may be interested in having all factor loadings as model parameters, whereas others may have only a few of them. Thus, unlike routine applications of regression analysis, there is no single way of assuming the appropriate parameters to be estimated without first considering the proposed model. Because the determination of model parameters is so important in setting up SEM models, it is discussed in detail next.

RULES FOR DETERMINING MODEL PARAMETERS

In order to correctly determine the parameters that can be uniquely estimated in a proposed SEM model, the following six rules will be used (cf. Bentler, 1995). It is important to note that when they are applied in practice, no distinction is made between the covariance and correlation of two (independent) variables (they are essentially considered equivalent in the sense of reflecting the degree of interrelationship between pairs of variables). The six rules are:

Rule 1. All variances of independent variables are model parameters. For example, in Fig. 6 most of the variances of independent variables are symbolized by asterisks associated with each error term. Error terms in a path diagram are generally attached to each dependent variable (e.g., E_1 to E_9). For the latent variables, the errors (or residuals) symbolize the structural regression disturbance terms. For example, the residual terms displayed in Fig. 3 (i.e., D_1 to D_3) encompass the effects on the corresponding dependent variable that are not accounted for by the influence of variables explicitly present in the model and impacting that dependent variable. It is important to note that all residual terms, whether attached to observed or latent variables, are unobserved entities because they cannot be measured and are independent variables by definition. Thus, by Rule 1, the variances of all residuals are, in general, model parameters. Of course, if there ever were a theory or hypothesis tested with a model that indicated that the variances of the independent variables (residual terms) were 0 or equal to a prespecified number, then Rule 1 would not apply.

Rule 2. All covariances between independent variables are model parameters (unless there is a theory or hypothesis being tested with the model that states that some of them are equal to 0 or equal to a given constant). In Fig. 6, the covariances between independent variables are the factor correlations symbolized by the two-way arrows connecting the three constructs. Note that this model does not hypothesize any correlation between observed variable residuals—there are no two-head arrows connecting any of the error terms—but other models may have one or more such correlations (e.g., see models presented in chap. 5).

Rule 3. All factor loadings connecting the latent variables with their indicators are model parameters (unless there is a theory or hypothesis tested with the model that states that some of them should be equal to 0 or equal to a given constant). In Fig. 6, these are the parameters denoted by the asterisks attached to the paths connecting each of the latent variables to its indicators.

Rule 4. All regression coefficients between observed or latent variables are model parameters (unless there is a theory or hypothesis tested with the model that states that some of them should be equal to 0 or equal to a given constant). For example, in Fig. 3 the regression coefficients are represented by the paths going from some latent variables and ending at other latent variables. It is important to note that Rule 3 can be considered a special case of Rule 4, after one observes that a factor loading can be conceived of as a regression coefficient (a slope) of the observed variable when regressed on the pertinent factor. However, in practice performing this particular regression is typically impossible because the factors are not observed variables to begin with and, hence, no individual measurements of them are available.

Rule 5. The variances and covariances between dependent variables and the covariances between dependent and independent variables are never model parameters. This is due to the fact that these variances and covariances are themselves explained in terms of the other model parameters. As can be seen in Fig. 6, there are no two-way arrows connecting dependent variables in the model or connecting dependent and independent variables.

Rule 6. For each latent variable included in a model, the metric of its latent scale needs to be set. The reason for this is that, unlike the observed variables in a model, there is no natural metric underlying the latent variables. In fact, unless this metric is defined, the scale of the latent variables will remain indeterminate. Subsequently, this leads to model-estimation problems and unidentified parameters and models (discussed later in this chapter). For any independent latent variable included in a proposed model, the metric can be fixed in one of two essentially equivalent ways. Either its variance is set equal to a constant (usually 1) or a path leaving the latent variable is set to a constant (also usually 1). For dependent latent variables, the metric fixing is done by setting a path leaving the latent variable equal a constant (typically 1). (The latest versions of some SEM programs, particularly LISREL, offer the option of automatically fixing the scales for both dependent and independent latent variable).

The reason that Rule 6 is needed stems from the fact that an application of Rule 1 on independent latent variables can produce a few redundant (i.e., unidentified) model parameters. For example, the variance and one of the paths emanating from each latent independent variable are redundant model parameters. The program is not able to estimate two redundant parameters as part of the model. As a result, one of them will be associated with an arbitrarily determined (and hence useless) estimate. This is because both parameters reflect the same aspect of a model (although in a different form), but cannot be uniquely estimated from the sampled data (i.e., they are not identifiable). Thus, a potentially infinite number of values can be estimated (all equally consistent with the data). Although the notion of identification is discussed extensively later in the book, note here that unidentified parameters can be made identified if one of them is considered equal to a constant (usually 1). This is the essence of Rule 6.

A Summary of Model Parameters in Fig. 6

Using these six rules, one can easily summarize the model parameters considered in Fig. 6. Following Rule 1, there are nine error term parameters (i.e., the variances of E_1 to E_9), as well as three factor variances (but they will be set to 1 shortly to follow Rule 6). Based on Rule 2, there are three factor covariance parameters. According to Rule 3, the nine factor loadings are model parameters. Rule 4 cannot be strictly applied in this model (unless it is considered a special case of Rule 3) because no regression-type relationships are assumed between any latent or observed variables. Rule 5 says that the relationships between the observed variables, which are the dependent variables of the model, are not parameters because they are supposed to be explained in terms of the actual model parameters. Similarly, the relationships between dependent and independent variables are not model parameters.

Rule 6 now implies that in order to fix the metric of the three latent variables one can set their variances to unity or fix (set to 1) one path leaving each of them. If a particularly good (fairly reliable) indicator of each latent variable is available, it is usually better to fix the scales of each latent variable by setting the path leading to that indicator to 1. Otherwise, it may be better to fix the scale of the latent variables by setting their variances to 1. It is also important to note that the paths leading from the nine error terms to their corresponding observed variables are not considered to be parameters, but instead are assumed to be equal to 1. For the latent variables in Fig. 6, one simply sets their variances equal to 1 (because all their loadings on the pertinent observed variables are already assumed to be model parameters). This setting the latent variance to 1 overrides the asterisks that would otherwise have to be attached to each latent variable circle in Fig. 6. Thus, applying all six rules, the model displayed in Fig. 6 has, altogether, 21 parameters to be estimated (i.e., nine error variances, plus nine factor loadings, plus three factor covariances). It is important to emphasize that the testing of any specific hypotheses in a model (e.g., setting all the indicator loadings on the Child intelligence factor to be the same value) places additional parameter restrictions and inevitably decreases the number of parameters to be estimated (discussed further in the next subsection). For example, if one assumes that the three loadings on the Child intelligence factor in Fig. 6 are equal to one another, it follows that they can be represented by a single model parameter. In this case, imposing a restriction on the model parameters decreases the number of parameters to be estimated to 19 (i.e., the number of model parameters decreases by two because the three factor loadings involved are not represented by three parameters anymore, but just by one).

Free, Fixed, and Constrained Parameters

There are three types of model parameters that are important in conducting SEM analyses—free, fixed, and constrained. All parameters that are determined based on the six rules are commonly referred to as *free model parameters* (unless a researcher imposes additional constraints on some of them; see later) and must be estimated by the SEM program. For example, in Fig. 6 asterisks were used to denote the free model parameters in the proposed factor analysis model. *Fixed parameters* have their value set to a given constant; such parameters are called fixed because they do not change value when the model is fit to the observed data. For example, in Fig. 6 the covariances (correlations) among the errors terms of the observed variables V_1 to V_9 are fixed parameters—the covariances are fixed to 0 in the proposed model. This is the reason there are no two-way arrows connecting any of the observed variable residuals. Moreover, following Rule 6 one may decide to set a factor loading or a latent variance equal to 1. In this case, the loading or variance also becomes a fixed parameter. Alternatively, a researcher may decide to fix other parameters that were initially conceived of as free parameters, which might represent substantively interesting hypotheses to be tested in a proposed model. Conversely, a researcher may decide to free some initially fixed parameters (after making sure, of course, that the model remains identified).

The other types of parameters are called constrained parameters (also sometimes referred to as restricted or restrained parameters). Models that include *constrained parameters* have parameters that are postulated to be equal to one another (but their value is not specified in advance as is that of the fixed parameters). Constrained parameters are typically included in a model if their restriction is derived from existing theory or represents a substantively interesting hypothesis tested in a proposed model. Thus, in a sense, constrained parameters can be considered to be somewhere between free and fixed parameters—constrained parameters are not completely free because they are set to follow some imposed restriction, yet their value can be anything as long as the restriction is preserved (but not locked at a particular constant as is a fixed parameter). It is for this reason that both free and constrained parameters are frequently referred to as model parameters.

For example, imagine a situation in which a researcher hypothesized that the factor loadings of the Parental dominance construct associated with the measures V_1, V_2, and V_3 in Fig. 6 were equal (referred to in the psychometric literature as a model with three tau-equivalent measures; e.g., Jöreskog, 1971). This hypothesis amounts to the assumption that the indicators measure the same latent variable in the same units of measurement. Thus, by using constrained parameters, a researcher can test the plausibility of this hypothesis. Of course, if constrained parameters are included in a model, their restriction should be derived from existing theory or formulated as a way to test substantively interesting hypotheses. Further discussion concerning the testing aspects of theoretically motivated hypotheses is provided in a later section of the book.

PARAMETER ESTIMATION

In any SEM model, parameters are estimated in such a way that the model becomes capable of "emulating" the analyzed sample covariance or correlation matrix (and as illustrated in chap. 6, in some circumstances the sample means). In order to clarify this feature of the estimation process, look again at the path diagram presented in Fig. 6 and the associated model definition Equations 1 in the previous section. As indicated in earlier discussions, the model represented by the path diagram and equations make certain assumptions about the relationships between the involved variables and, hence, have specific implications for their variances and covariances. It turns out that these implications can be worked out using a few simple relations that govern the variances and covariances of linear combinations of variables. These relations are illustrated next as the four laws of variances and covariance, which follow straightforwardly from the formal definition of variances and covariances (e.g., Hays, 1994).

The Four Laws for Variances and Covariances

Denote variance by the label Var and covariance by Cov. For a random variable X (e.g., an intelligence test score), the first law is stated as follows:

Law 1.

$$\text{Cov}(X,X) = \text{Var}(X).$$

Law 1 simply says that the covariance of a variable with itself is that variable's variance. This is an intuitively very clear result that is a direct consequence of the definition of variance and covariance. (It can be readily seen in action by looking at the formula for the estimation of variance and observing that it results from the formula for estimating covariance when the two variables involved coincide; e.g., Hays, 1994.)

The second law allows one to find the covariance of two linear combinations of variables. Assume that X, Y, Z, and U are four random variables (e.g., those denoting the scores on tests of depression, social support, intelligence, and a person's age; see Equation 2 in the section Rules for Determining Model Parameters). Suppose that a, b, c, and d are four constants. Then the following relationship holds:

Law 2.

$$\text{Cov}(aX + bY, cZ + dU) = ac\,\text{Cov}(X,Z) + ad\,\text{Cov}(X,U) + bc\,\text{Cov}(Y,Z) + bd\,\text{Cov}(Y,U).$$

This law is easy to remember because it is quite similar to the rule of disclosing brackets used in elementary algebra. Indeed, to apply Law 2 simply determine each resulting product of constants and attach the covariance of their pertinent variables. Note that the right-hand side of the law simplifies markedly if some of the variables are uncorrelated (i.e., one or more of the involved covariances is equal to 0).[2]

Using Laws 1 and 2 (and the knowledge that $Cov(X,Y) = Cov(Y,X)$, recalling that covariance does not depend on the order of the variables), one obtains the next equation, which, due to its importance for the remainder of the book, is formulated as a separate law:

Law 3.

$$Var(aX + bY) = Cov(aX + bY, aX + bY)$$
$$= a^2 Cov(X,X) + b^2 Cov(Y,Y) + ab Cov(X,Y) + ab Cov(X,Y),$$

or simply

$$= a^2 Var(X) + b^2 Var(Y) + 2ab Cov(X,Y).$$

A special case of Law 3 that is used often in this book involves uncorrelated variables X and Y (i.e., $Cov(X,Y) = 0$), and, as such, it is formulated as a separate law:

Law 4. If X and Y are uncorrelated, then

$$Var(aX + bY) = a^2 Var(X) + b^2 Var(Y).$$

It is important to note that there are no restrictions in Laws 2, 3, and 4 on the values of the constants a, b, c, and d (they can even take on the values 0 or 1). In addition, these laws generalize straightforwardly to the case of linear combinations with more than two variables.

---

[2]Law 2 reveals the rationale behind the six rules (see the section Rules for Determining Model Parameters) for determining the parameters for any model once the definition equations are written down. Law 2 states that the covariance of any pair of observed measures (or any measure's variance) is a function of the covariances or variances of the variables or both, and the weights by which these variables are multipled and then summed up in the equations of these measures. The variables mentioned are the pertinent independent variables of the model (their analogs in Law 2 are X, Y, Z, and U). The weights mentioned are the pertinent factor loadings or regression coefficients in the model (their analogs in Law 2 are the constants a, b, c, and d. Therefore, the parameters of any SEM model are (a) the variances and covariances of the independent variables, and (b) the factor loadings or regression coefficients (unless there is a theory or hypothesis tested within the model that states that some of them are equal to constants, in which case the parameters are the remaining quantities in (a) and (b)).

Model Implications and Reproduced Covariance Matrix

As mentioned earlier in this section, any proposed model has certain implications for the variances and covariances (and the means, if considered) of the involved observed variables. In order to see these implications, the four laws for variances and covariances must be used. For example, consider the first two manifest variables V_1 and V_2 presented in Equations 1 (see the section Rules for Determining Model Parameters and Fig. 6). Because both variables load on the same latent factor F_1, we obtain the following equality directly from Law 2:

$$
\begin{aligned}
\text{Cov}(V_1, V_2) &= \text{Cov}(\lambda_1 F_1 + E_1, \lambda_2 F_1 + E_2) \\
&= \lambda_1 \lambda_2 \, \text{Cov}(F_1, F_1) + \lambda_1 \, \text{Cov}(F_1, E_2) \\
&\quad + \lambda_2 \, \text{Cov}(E_1, F_1) + \text{Cov}(E_1, E_2) \\
&= \lambda_1 \lambda_2 \, \text{Cov}(F_1, F_1) \\
&= \lambda_1 \lambda_2 \, \text{Var}(F_1, F_1) \\
&= \lambda_1 \lambda_2 .
\end{aligned}
\tag{3}
$$

To obtain Equation 3, the following two facts (based on the proposed model in Fig. 6) are also used. First, the covariance of the residuals E_1 and E_2, and the covariance of each of them with the factor F_1 are equal to 0 (i.e., in Fig. 6 there are no two-headed arrows connecting the residuals or any of them with F_1). Second, the variance of F_1 has been set equal to 1 according to Rule 6 (i.e., $\text{Var}(F_1, F_1) = 1$).

Similarly, using Law 2, the covariance between the observed variables V_1 and V_4 (each loading on a different factor) is determined as follows:

$$
\begin{aligned}
\text{Cov}(V_1, V_4) &= \text{Cov}(\lambda_1 F_1 + E_1, \lambda_4 F_2 + E_4) \\
&= \lambda_1 \lambda_4 \, \text{Cov}(F_1, F_2) + \lambda_1 \, \text{Cov}(F_1, E_4) \\
&\quad + \lambda_4 \, \text{Cov}(E_1, F_2) + \text{Cov}(E_1, E_4) \\
&= \lambda_1 \lambda_4 \phi_{21},
\end{aligned}
\tag{4}
$$

where ϕ_{21} (Greek letter *phi*) is used to denote the covariance between the factors F_1 and F_2.

Finally, the variance of the observed variable V_1 is determined, using Law 4 and the previously stated facts, as:

$$
\begin{aligned}
\text{Var}(V_1) &= \text{Cov}(\lambda_1 F_1 + E_1, \lambda_1 F_1 + E_1) \\
&= \lambda_1^2 \, \text{Cov}(F_1, F_1) + \lambda_1 \, \text{Cov}(F_1, E_1) + \lambda_1 \, \text{Cov}(E_1, F_1) + \text{Cov}(E_1, E_1) \\
&= \lambda_1^2 \, \text{Var}(F_1) + \text{Var}(E_1) \\
&= \lambda_1^2 + \theta_1,
\end{aligned}
\tag{5}
$$

where θ_1 (Greek letter *theta*) is used to symbolize the variance of the residual E_1.

If this process were continued for every combination of p observed variables (i.e., V_1 to V_9), the result would be the determination of every element of a variance–covariance matrix. This matrix can be denoted by Σ (the Greek letter *sigma*) and is generally referred to as the reproduced (or model-implied) covariance matrix. Because the Σ matrix is symmetric (the elements above the main diagonal are identical to the elements below the main diagonal), it has altogether $p(p + 1)/2 = 9(10)/2 = 45$ nonredundant elements. Interestingly, the number of nonredundant elements will also be used later in this chapter to determine the degrees of freedom of a proposed model.

Thus, using Laws 1 through 4 for the proposed model in Fig. 6, the following reproduced covariance matrix Σ is determined (displaying only its diagonal and lower off-diagonal elements):

$$
\Sigma =
\begin{array}{llllllllll}
\lambda_1^2 + \theta_1 \\
\lambda_1\lambda_2 & \lambda_2^2 + \theta_2 \\
\lambda_1\lambda_3 & \lambda_2\lambda_3 & \lambda_3^2 + \theta_3 \\
\lambda_1\lambda_4\phi_{21} & \lambda_2\lambda_4\phi_{21} & \lambda_3\lambda_4\phi_{21} & \lambda_4^2 + \theta_4 \\
\lambda_1\lambda_5\phi_{21} & \lambda_2\lambda_5\phi_{21} & \lambda_3\lambda_5\phi_{21} & \lambda_4\lambda_5 & \lambda_5^2 + \theta_5 \\
\lambda_1\lambda_6\phi_{21} & \lambda_2\lambda_6\phi_{21} & \lambda_3\lambda_6\phi_{21} & \lambda_4\lambda_6 & \lambda_5\lambda_6 & \lambda_6^2 + \theta_6 \\
\lambda_1\lambda_7\phi_{21} & \lambda_2\lambda_7\phi_{31} & \lambda_3\lambda_7\phi_{31} & \lambda_4\lambda_7\phi_{32} & \lambda_5\lambda_7\phi_{32} & \lambda_6\lambda_7\phi_{32} & \lambda_7^2 + \theta_7 \\
\lambda_1\lambda_8\phi_{31} & \lambda_2\lambda_8\phi_{31} & \lambda_3\lambda_8\phi_{31} & \lambda_4\lambda_8\phi_{32} & \lambda_5\lambda_8\phi_{32} & \lambda_6\lambda_8\phi_{32} & \lambda_7\lambda_8 & \lambda_8^2 + \theta_8 \\
\lambda_1\lambda_9\phi_{31} & \lambda_2\lambda_9\phi_{31} & \lambda_3\lambda_9\phi_{31} & \lambda_4\lambda_9\phi_{32} & \lambda_5\lambda_9\phi_{32} & \lambda_6\lambda_9\phi_{32} & \lambda_7\lambda_9 & \lambda_8\lambda_9 & \lambda_9^2 + \theta_9.
\end{array}
$$

It is important to note that the elements of Σ are all functions of model parameters. In addition, each element of Σ has as a counterpart a corresponding numerical element (entry) in the observed sample covariance matrix obtained for the nine observed variables considered. Assuming that the observed covariance matrix (denoted by S) was as follows:

$$
S =
\begin{array}{lllllllll}
1.01 \\
0.32 & 1.50 \\
0.43 & 0.40 & 1.22 \\
0.38 & 0.25 & 0.33 & 1.13 \\
0.30 & 0.20 & 0.30 & 0.70 & 1.06 \\
0.33 & 0.22 & 0.38 & 0.72 & 0.69 & 1.12 \\
0.20 & 0.08 & 0.07 & 0.20 & 0.27 & 0.20 & 1.30 \\
0.33 & 0.19 & 0.22 & 0.09 & 0.22 & 0.12 & 0.69 & 1.07 \\
0.52 & 0.27 & 0.36 & 0.33 & 0.37 & 0.29 & 0.50 & 0.62 & 1.16,
\end{array}
$$

then the top element value of S (i.e., 1.01) corresponds to $\lambda_1^2 + \theta_1$ in the reproduced matrix Σ. Similarly, the counterpart of the element in the sixth row and fourth column of S (i.e., 0.78) is $\lambda_4\lambda_6$ in the Σ matrix.

Now imagine setting the counterpart elements of S and Σ equal to one another, from the top-left corner of S to its bottom-right corner. That is, according to the proposed model displayed in Fig. 6, set $1.01 = \lambda_1^2 + \theta_1$, then $0.32 = \lambda_1\lambda_2$, and so on until for the last elements, $1.16 = \lambda_9^2 + \theta_9$ is set. As a result of this equality setting, a system of 45 equations (i.e., the number of nonredundant elements) with as many unknowns as there are model parameters (i.e., 21 asterisks in Fig. 6) is generated. Thus, one can conceive of the process of fitting a structural equation model as a way of solving a system of equations. For each equation, its left-hand side is a subsequent numerical entry of the sample covariance matrix S, whereas its right-hand side is the corresponding expression of model parameters defined in the matrix Σ. Hence, fitting a structural equation model is conceptually equivalent to solving this system of equations obtained according to the proposed model in an optimal way (discussed in the next section).

The preceding discussion also demonstrates that the model presented in Fig. 6, as does any structural equation model, implies a specific structuring of the elements of the covariance matrix reproduced by the model in terms of specific expressions (functions) of unknown model parameters. Therefore, if certain values for the parameters were entered into these functions, one would obtain a covariance matrix that has numbers as elements. In fact, the process of fitting a model to data with SEM programs can be thought of as repeated insertion of appropriate values for the parameters in the matrix Σ until a certain optimality criterion (discussed in the next section), in terms of its proximity to the matrix S, is satisfied.

Every available SEM program has built into its memory the exact way in which these functions of model parameters in Σ can be obtained. Although for ease of computation most programs make use of matrix algebra, the programs in effect determine each of the expressions presented in these 45 equations (see Marcoulides & Hershberger, 1997 for further discussion). Fortunately, this occurs quite automatically once the user has communicated to the program the model parameters (and a few other related details discussed in the next chapter).

How Good Is the Proposed Model?

The previous subsection illustrated how a proposed SEM model leads to a reproduced covariance matrix Σ that is fit to the observed sample covariance matrix S. Now it would seem that the next logical question is, "How can one measure or evaluate the extent to which the matrices S and Σ differ?" As it turns out, this question is a particularly important question in SEM because it actually permits one to evaluate the goodness of fit of the model. Indeed, if the difference between S and Σ is small, then one can conclude that the model represents the observed data reasonably well. On the other hand, if the difference is large, one can conclude that the proposed model is not consistent with the observed data. There are at least two reasons for such inconsistencies: (a) the proposed model may be deficient, in the sense that it is not capable of emulating the analyzed matrix even with most favorable parameter values or (b) the data may not be good. Thus, in order to proceed with assessing model fit a method is needed for evaluating the degree to which the reproduced matrix Σ differs from the sample covariance matrix S.

In order to clarify this method, a new concept is introduced, that of distance between matrices. Obviously, if the values to be compared were scalars (single numbers) a simple subtraction of one from the other (and possibly taking the absolute value of the resulting difference) would suffice to evaluate the distance between them. However, this cannot be done directly with the two matrices S and Σ. Subtracting the matrix S from the matrix Σ does not result in a single number; rather, a matrix of differences is obtained.

Fortunately, there are some meaningful ways to assess the distance between two matrices. And, interestingly, the resulting distance measure ends up being a single number that is easier to interpret. Perhaps the simplest way to obtain this single number involves taking the sum of the squares of the differences between the corresponding elements of the two matrices. Other more complicated ways involve the multiplication of these squares with some appropriately chosen weights and then taking their sum (discussed later). In either case, the single number represents a sort of generalized distance measure between the two matrices considered. The bigger the number, the more different the matrices are, and the smaller the number, the more similar the matrices. Because in SEM this number results after comparing the elements of S with those of the model-implied covariance matrix Σ, the generalized distance is a function of the model parameters as well as the elements of the observed variances and covariances. Therefore, it is customary to refer to the relationship between the matrix distance, on the one hand, and the model parameters and S, on the other hand, as a *fit function*, which is typically denoted by F. Because it equals the distance between two matrices, F is always equal to a positive value or 0. Whenever the value of F is 0, then the two matrices considered are identical.

It turns out that depending on how the relationship between matrix distances, model parameters, and the elements of S are defined, several fit functions may result. The corresponding fit functions, along with each characteristic method of parameter estimation, are discussed next.

Methods of Parameter Estimation

There are four main estimation methods and types of fit functions that are used in most structural modeling programs: unweighted least squares, maximum likelihood, generalized least squares, and asymptotically distribution free (often called weighted least squares). The application of each estimation method is based on the minimization of a corresponding fit function.

The unweighted least squares (ULS) method uses as a fit function (denoted by F_{uls}), the simple unweighted sum of squared differences between the corresponding elements of S and the model reproduced covariance matrix Σ. Accordingly, estimates are chosen for the model parameters when F_{uls} attains its smallest value. The ULS method is usually used in practice when similar scales of measurement underlie the analyzed variables.

The other three estimation methods are based on the same sum of squares, but after specific weights have been used to multiply each of the squares. The maximum likelihood (ML) and the generalized least squares (GLS) methods are used when the observed data are normally distributed. The assumption of normality is quite frequently made in multivariate analyses. The assumption of normality can be examined using any general-purpose statistical package (e.g., SAS or SPSS) or even using EQS and PRELIS (a subprogram in the LISREL program, labeled the precursor of LISREL; Jöreskog & Sörbom, 1993c). The simplest way to examine univariate normality is to consider skewness and kurtosis, but statistical tests are also available for this purpose. Skewness is an index that reflects the symmetry of a univariate distribution. Kurtosis has to do with the shape of the distribution in terms of its peakedness. Under normality, the univariate skewness and kurtosis coefficients should be 0. There is also a measure of multivariate kurtosis called Mardia's multivariate kurtosis coefficient and its normalized estimate (Bentler, 1995). Mardia's multivariate kurtosis coefficient measures the extent to which the multivariate distribution of all observed variables has tails that differ from the ones characteristic of the normal distribution (with the same component means, variances, and covariances). If the distribution deviates only marginally from the normal, Mardia's coefficient will be close to 0 and its normalized estimate probably nonsignificant.

Although it may be quite likely that the assumption of multivariate normality is met if all observed variables are individually normally distributed, it is preferable to also examine bivariate normality. If the observations are from a multivariate normal distribution, each bivariate distribution should also be normal. One method for examining normality involves looking at the scatter plots between all pairs of analyzed variables to ensure that they have (at least approximately) cigar-shaped forms (e.g., Tabachnik & Fidell, 1999). A somewhat more formal method for judging bivariate normality is based on a plot of the chi-square percentiles and the mean distance measure of individual observations. If the distribution is normal, the plot of the chi-square percentiles and the mean distance measure should resemble a straight line (for complete details on assessing multivariate normality see Marcoulides & Hershberger, 1997, pp. 48–52).

In recent years, research has shown that the maximum likelihood (ML) method can also be employed with minor deviations from normality (e.g., Bollen, 1989; Jöreskog & Sörbom, 1993b; see also Raykov & Widaman, 1995). For this reason (as well as the fact that this is an introductory book) only the ML method is presented in this book. Broadly speaking, the ML method determines estimates for the model parameters that maximize the likelihood of observing the available data if one were to collect data from the same population again. This maximization is achieved by selecting (using a numerical search procedure across the space of all possible parameter values) the model parameters in such a way that they minimize the fit function described earlier (denoted as F_{ML}).

With more serious deviations from normality, the asymptotically distribution free (or weighted least squares) method can be used as long as the size of the analyzed sample is large. There is no doubt that sample size plays an important role in almost every statistical technique applied in practice. Although there is universal agreement among researchers that the larger the sample the more stable the parameter estimates, there is no agreement as to what constitutes large. This topic has received a considerable amount of attention in the literature, but no easily applicable and clear-cut general rules of thumb have been proposed. To give only an idea of the issue involved, a cautious attempt at a rule of thumb suggests that the sample size should always be more than 10 times the number of free model parameters (cf. Bentler, 1995; Hu, Bentler, & Kano, 1992). Otherwise, the results from the asymptotically distribution free (ADF) method should generally not be trusted. If the sample size is smaller, researchers are encouraged to use the Satorra–Bentler robust method of parameter estimation (a special type of ADF method), which is available in the EQS program.

Another alternative to dealing with nonnormal data is to make the data more normal looking by introducing some normalizing transformation on the raw data. Once the data have been transformed, normal theory analysis can be carried out. In general, transformations are simply a reexpression of the data in different units of magnitude. Numerous transformation methods have been proposed in the literature. The most popular transformations included in most general-purpose statistical packages are power transformations (e.g., squaring each data point), square root transformations, reciprocal transformations, and logarithmic transformations. Finally, one may also want to consider data on other measures of the constructs involved in the proposed model if such are readily available.

With data stemming from designs with only a few possible response categories, the asymptotically distribution free method is usually used with polychoric or polyserial correlations. For example, suppose a questionnaire includes the item, "How satisfied are you with your recent car purchase?" with response categories labeled, "Very satisfied," "Somewhat satisfied," and "Not satisfied." A considerable amount of research has shown that ignoring the categorical attributes of data obtained from items like these can lead to biased SEM results. For this reason, researchers suggest using the polychoric-correlation coefficient (for assessing the degree of association between ordinal variables) and the polyserial-correlation coefficient (for assessing the degree of association between an ordinal variable and a continuous variable). Fortunately, some research has shown that when there are five or more response categories (and the distribution of data looks normal) the problems from disregarding the categorical nature of responses are likely to be minimized (Rigdon, 1998). Thus, once again, examining the distributions of the data becomes essential.

From a statistical perspective, all four parameter estimation methods lead to consistent estimates. Consistency is a desirable feature that insures that with increasing sample size the estimates converge to the true population parameter values. With large samples, the estimates obtained from the maximum likelihood, generalized least squares, or asymptotically distribution free methods also possess the additional important property of being normally distributed around their population counterparts. Moreover, with large samples these three methods yield efficient estimates (i.e., having the smallest possible variances), hence representing the most precise (least unstable) estimates of the parameters of interest.

Iterative Estimation of Model Parameters

The final question now is, "Using any estimation method, how does one actually estimate the parameter values for the proposed model in order to render the covariance matrix Σ as close as possible to the observed covariance matrix S?" It turns out that in order to answer this question one must resort to the numerical routines implemented in available SEM programs. The goal of the numerical routines is basically to minimize the corresponding fit function. These numerical routines proceed in a consecutive (iterative) manner selecting values for model parameters in such a way that at each step the distance between Σ and S is reduced until no further improvement in the fit function can be achieved (i.e., there is no further decrease in the generalized distance between Σ and S).

The iterative process described earlier starts with initial estimates of the parameters. Fortunately, the initial estimates are automatically calculated by the SEM programs (although users can provide their own initial estimates if they so choose). The iteration process terminates (or converges) if at some step the fit function F does not change by more than a very small amount (in most cases .000001, although even this value can be changed by the program user). Thus, the iterative process of parameter estimation converges when there is practically no further improvement in the distance between the two matrices Σ and S. The numerical values for the parameters obtained at the final iteration step are considered the final solution values and represent the required estimates of the model parameters. It is important to note that in order for a set of parameter estimates to be meaningful, it is necessary that the iterative process converge (terminate) to a final solution. If convergence does not occur, a warning sign is issued by SEM programs (which is easily spotted in the output), and the results are meaningless (beyond being useful for tracking down the problem of the lack of convergence).

All converged solutions also provide a measure of the sampling variability for each obtained parameter estimate, called the *standard error*. The magnitude of the standard errors indicates how stable the parameter estimate is if repeated sampling were carried out from the population of interest. With plausible models (see further discussion later), the standard errors are used to compute t values, which provide information about the statistical significance of the parameter estimate. The t values are computed as the ratio of the parameter estimate to its standard error. If for a free parameter, its t value is greater than $+2$ or less than -2, the parameter is referred to as significant and can be considered distinct from 0 in the population. Conversely, if its t value is between ± 2, the parameter is nonsignificant and can be considered 0 in the population. Furthermore, based on the large-sample normality feature, adding twice the standard error to and subtracting twice the standard error from the parameter estimate yields a confidence interval (at the 95% confidence level) for that parameter. This confidence interval represents a range of plausible values for the parameter in the population and can be conventionally used to test hypotheses about a prespecified value of the parameter in the population. In particular, if the computed interval covers the prespecified value, the hypothesis is retained (at significance level .05); otherwise, it is rejected. Moreover, the width of the interval permits one to assess the precision of estimation of the parameter. Wider confidence intervals are associated with lower precision (and larger standard errors), and narrower confidence intervals with higher precision of estimation (and smaller standard errors). In fact, these features of the standard errors as measures of sampling variability of the parameter estimates make them quite useful (along with a number of model goodness-of-fit indices discussed in the next section) for assessing the goodness of fit of the model.

It is also important to note that another reason that the numerical fit-function minimization procedure may not converge could be that the proposed model may simply be inadequate for the analyzed data, or may contain one or more unidentified parameters. Unidentified parameters do not posses unique estimates—unlike identified parameters—even if one has gathered data on the analyzed variables from the whole population and calculated their population covariance matrix to which the model is subsequently fitted. It is therefore of utmost importance to deal only with identifiable parameters, and thus have an identified model. Due to its great importance in conducting SEM analyses, the issue of parameter and model identification is discussed next.

PARAMETER AND MODEL IDENTIFICATION

A model parameter is unidentified if there is not enough empirical information to allow its unique estimation. As such, any estimate of an unidentified parameter computed by a SEM program is arbitrary and should not be relied on. A model containing at least one unidentified parameter cannot generally be relied on either, even though some parts of it may represent a useful approximation to the studied phenomenon. Because an unidentified model is generally useless in practice (even though it may be useful in some theoretical discussions), one must ensure the identification of a model by following some general guidelines.

What Does It Mean to Have an Unidentified Parameter?

In simple terms, having an unidentified parameter implies that it is impossible to compute a reasonable estimate of it. For example, suppose one is considering the equation $a + b = 10$, and is faced with the task of finding unique values for two unknown constants a and b in order to satisfy the equation. One solution could be $a = 5, b = 5$, and another could be $a = 1$, $b = 9$. Obviously, there is no way in which one can determine unique values for a and b because it involves a single equation with two unknowns. As such, there are an infinite number of solutions to the equation because there are more unknown values (two parameters) than known values (one equation). Thus, the model represented by this equation is considered underidentified (or unidentified), and any estimates obtained are meaningless. As will become clear later, the most desirable condition to encounter in SEM is to have more equations than are needed to obtain unique solutions for the estimates. This condition is called overidentification.

A similar identification problem occurs when the only information available is that the product of two unknown constants x and y is equal to 55. Knowing only that $xy = 55$ does not provide sufficient information to come up with unique estimates of either x or y. Of course, one could choose a value of x in a completely arbitrary manner (except, of course, 0) and then take $y = 55/x$ to provide one solution. Because there are an infinite number of solutions for x and y, neither can be identified until some further information is provided (e.g., a preselected value for one of them). As it turns out, this product example is relevant in the context of the discussion of Rule 6 for determining model parameters. If neither the variance of a latent variable nor a path going from it are fixed to a constant, then this example demonstrates how the variance and any factor loading become entangled in their product and hence are unidentified.

Although the two numerical examples presented were deliberately rather simple, they nonetheless illustrate how similar problems can occur in the context of SEM models. Recall from earlier sections that SEM can be thought of as an approach to solving, in an optimal way, a system of equations—those relating the elements of the sample covariance matrix S with their counterparts in the model reproduced covariance matrix Σ. It is possible then, depending on the model, that for some of its parameters the system may have infinitely many solutions. Clearly, such a situation is not desirable given the fact that SEM models attempt to determine what the parameter values look like in the population. Only identified models and parameter estimates can provide this information. A straightforward way to determine if a model is unidentified is presented in the next section.

A Necessary Condition for Model Identification

The parallel between SEM and solving a system of equations is also useful for remembering a simple but necessary condition of model identification. Specifically, if the system of equations relating the elements of the sample covariance matrix S with their counterparts in the model-reproduced covariance matrix Σ has more unknowns than equations, then the model will definitely be unidentified. This is because the system of equations contains more parameters than could possibly be uniquely solved. Although this condition is easy to check, it should be noted that it is only a necessary condition for model identification; it is not a sufficient condition. In other words, having less unknown parameters than equations in that system does not guarantee that a model is identified. However, if a model is identified, the condition must hold (i.e., there will be fewer parameters than nonredundant elements of the covariance matrix S).

To check this necessary condition, one simply counts the number of parameters in the model (i.e., the number of asterisks included in the proposed model path diagram), and subtracts the value from the number of nonredundant elements in the sample covariance matrix S (i.e., $p(p+1)/2$, where p is the number of observed variables). The resulting difference,

$$\frac{p(p+1)}{2} - \text{(Number of model parameters)}, \tag{6}$$

is referred to as the *degrees of freedom* of the proposed model and usually denoted as *df* (some researchers also refer to this check as the *t* rule). If the difference is positive, the necessary condition for model identification is fulfilled. If Equation 6 is 0, then the degrees of freedom are 0 and the model is called saturated or just-identified. *Saturated models* have as many parameters as there are nonredundant elements in the covariance matrix. As it turns out, there is no way one can really test or confirm the plausibility of a saturated (just-identified) model. This is because saturated models will always fit the data perfectly. Because one of the primary reasons for conducting a SEM analysis is to test the fit of a proposed model, at a minimum, this requires positive degrees of freedom (i.e., more unique elements in the data covariance matrix than parameters that need to be estimated). Thus, if Equation 6 is negative, the model is definitely unidentified (and some SEM programs may alert the user of an identification problem by generating a statement that the degrees of freedom in the model are negative). As indicated, having positive degrees of freedom in a proposed model is a necessary but not a sufficient condition for identification. There are situations in which the degrees of freedom for a proposed model are quite high and yet some of the parameters remain underidentified. Model identification is an enormously complex issue that requires careful consideration and handling.

How to Deal with Unidentified Parameters in Practice

If a proposed model is carefully conceptualized, the likelihood of unidentified parameters will usually be minimized. In particular, using Rules 1 to 6 will most likely ensure that the proposed model is identified. However, if a model is found to be unidentified, a first step toward identification is to see if all its parameters have been correctly determined or whether all the latent variables have their scales fixed. In many instances, a SEM program will signal an identification problem with an error message and even correctly point to the unidentified parameter. Unfortunately, in some instances it may point to the wrong parameter. Hence, the best strategy is for the researcher to accurately locate the unidentified parameters in a model before trying to correct the problem. One possible way of dealing with unidentified parameters is to impose appropriate, substantively plausible constraints on them or on some functions of them (and possibly on other parameters). Of course, because this way of attempting model identification may not always work, often either a completely new model may have to be contemplated (one that does not contain this parameter), or a new study and data-collection process may have to be designed.

MODEL-TESTING AND -FIT EVALUATION

SEM methodology offers researchers a method for the quantification and testing of theories. The theories are represented in models that describe and explain the phenomena under investigation. As described previously, an essential requirement for all such models is that they be identified. Another requirement (perhaps of equal importance) is that researchers consider for study only those models that are attached to some substantive considerations and represent plausible means of data description and explanation.

SEM methodology offers a number of inferential and descriptive indices that reflect the extent to which a model can be considered an acceptable means of data representation. Using them together with substantive considerations allows one to make a decision if a given model should reasonably be rejected as a means of data explanation or should be relied on (to some extent). The topic of *model-fit evaluation* (the evaluation of the extent to which a model fits an analyzed data set) in SEM is very broad and complex. Due to the introductory nature of this book, this section provides a relatively brief discussion of the topic, which can be considered a minimalist's scheme for carrying out model evaluation. For further elaboration on these issues, we refer the reader to Bentler (1995), Bollen (1989), Byrne (1998), Jöreskog and Sörbom (1993a, 1993b), Marcoulides and Hershberger (1997), Marcoulides and Schumacker (1996), or Schumacker and Lomax (1996).

Substantive Considerations

A major aspect of model-fit evaluation involves the issue of the substantive considerations of a model. Specifically, all models considered in research should be conceptualized according to the latest knowledge about the phenomenon under consideration. This knowledge is usually obtained after an extensive study of the pertinent literature. As such, all proposed models should try to embody in appropriate ways the results available from previous studies. Of course, if a researcher wishes to study a model that contradicts past theories, then alternative models with various new restrictions or relationships between involved variables can also be tested. But the initial conceptualization of any proposed model can only come after an informed study of the phenomenon under consideration.

Regardless of whether a researcher proposes a model that supports or contradicts past knowledge, the advantages of SEM methodology can only be used with variables that have been validly and reliably assessed. Even the most intricate and sophisticated models are of no use if the variables included in the model are poorly assessed. A model cannot do more than what is contained in the data themselves. If the data are poor, in the sense of reflecting substantial unreliability in the analyzed data, the results will be poor, regardless of the quality of the model.

Providing an extensive discussion of the various ways of ensuring the measurement properties of variables included in proposed models is beyond the scope of this introductory book. These issues are usually addressed at length in books dealing specifically with psychometrics and measurement theory (e.g., Allen & Yen, 1979; Crocker & Algina, 1986; Suen, 1990). The present book assumes that the researcher has sufficient knowledge of how to organize a reliable and valid assessment of the variables included in a proposed model.

Model Evaluation and the True Model

Before particular measures of model fit are discussed, a word of warning is in order. Even if all possible fit indices point to an acceptable model, one can never claim to have found the true model that has generated the analyzed data (of course, the cases in which data is simulated according to a preset known model are excluded from consideration). This fact is rooted in another specificity of the SEM methodology that is different from classical modeling approaches. Whereas classical methodology is typically interested in rejecting null hypotheses (because the substantive conjecture is usually reflected in the alternative hypotheses of difference or change), SEM is most concerned with finding a model that does not contradict the data. That is, in an empirical session of SEM, one is typically interested in retaining the proposed model whose validity is the essence of the null hypothesis. In other words, statistically speaking, when using SEM methodology one is usually interested in not rejecting the null hypothesis.

However, recall from introductory statistics that not rejecting a null hypothesis does not mean that it is true. Similarly, because model testing in SEM involves testing the null hypothesis that the model is capable of perfectly reproducing (with certain values of its unknown parameters) the analyzed matrix of observed variable interrelationship indices, not rejecting a fitted model does not imply that it is the true model. In fact, it may well be that the model is not correctly specified (i.e., wrong), yet due to sampling errors (that are in effect when gathering the data) it appears plausible. Similarly, just because a model fits a data set well does not mean that it is the only model that fits the data well. There can be a plethora of other models that fit the data equally well or even better. In fact, there can be a number (possibly very many; see Raykov & Marcoulides, in press) of equivalent models that fit the data just as well as a model under consideration. Unfortunately, at present there is no statistical means for discriminating among these equivalent models—especially when the issue is choosing one (or more) of them for further consideration or interpretation. Which one of these models is better and which one is wrong can only be decided on the basis of a sound body of knowledge about the studied phenomenon. This is partly the reason that substantive considerations are important in model-fit evaluation. In addition, one can also evaluate the validity of a proposed model by conducting replication studies. The value of a proposed model is greatly enhanced if the same model can be replicated in new samples from the same population.

Parameter Estimate Signs, Magnitude, and Standard Errors

It is worth reiterating at this point that one cannot interpret a model solution provided by a structural modeling program if the numerical minimization routine does not converge (i.e., has not ended after a finite number of iterations). If the routine does not terminate, one cannot trust the output of any program for purposes of solution interpretation (although the solution may provide information for tracking down the reasons for the lack of convergence).

For a model to be considered further for fit evaluation, the parameter estimates in the final solution of the minimization procedure should have the right sign and magnitude as predicted or expected by theory. In addition, the standard errors associated with each of the parameter estimates should not be excessively large. If a standard error of a parameter estimate is large (especially when compared to other parameter standard errors), the model does not provide reliable information and should be interpreted with great caution (and, if possible, the reasons for this finding clarified).

Goodness-of-Fit Indices

Chi-Square Value.   Evaluation of model fit is typically carried out on the basis of an inferential goodness-of-fit index as well as a number of other descriptive indices. The inferential index is called a chi-square value. It represents a test statistic of the goodness of fit of the model, and it is used when testing the null hypothesis that the model fits the analyzed covariance matrix perfectly. This test statistic is defined as

$$T = (N - 1) \, F_{min}, \tag{7}$$

where N is the sample size and F_{min} is the computed minimal value of the fit function for the parameter estimation method used (e.g., ML, GLS, or ADF).

The name chi-square value derives from the fact that with large samples the distribution of T approaches a (central) chi-square distribution, if the model is correct and fitted to the covariance matrix S. The degrees of freedom of this distribution are equal to those of the model. As mentioned previously, the degrees of freedom are determined by using the formula $df = (p(p+1)/2) - q$, where p is the number of observed variables of the model and q is the number of model parameters (see Equation 6 in the section Parameter and Model Identification).

When the proposed model is fit to the data using a SEM program, the program will judge the obtained chi-square value T in relation to the model's degrees of freedom, and output its associated p value. The p value can be examined and compared with a preset significance level (often .05) in order to test the hypothesis that the model is capable of (exactly) reproducing the analyzed matrix of variable relationship indices. Hence, following the statistical hypothesis-testing tradition, one may consider the rejection of a model when its p value is smaller than the preset significance value (e.g., .05), and the retention of the model if this value is higher than the preset significance.

Although this way of looking at statistical inference in SEM may appear to be the reverse of the one used in the framework of traditional hypothesis testing, it turns out that, at least from a philosophy-of-science perspective, the two are compatible. Indeed, following Popperian logic (Popper, 1962), one's interest lies in rejecting models rather than confirming them because there is no scientific way of proving the validity of a proposed model. That is, in practice no SEM model can be proved to be the true model (see discussion in previous section).

In this context, it is also important to note that, in general, there is a preference for dealing with models that have a large number of degrees of freedom. This is because an intuitive meaning of the notion of degree of freedom is a dimension along which the model can be rejected. Thus, the more degrees of freedom, the more dimensions there are along which one can reject the model, and hence the higher the likelihood of rejecting it when it is tested against the data. This is a desirable feature of the testing process because, according to Popperian logic, empirical science can only disconfirm not confirm models. This view also entails that, if one has two models that are plausible descriptions of a studied phenomenon, the one with more degrees of freedom is a stronger candidate for consideration as a means of data description and explanation. This is because with more degrees of freedom the model has withstood a greater chance of being rejected; if the model was not rejected, the results are more trustworthy. This is essentially the conceptual basis of the parsimony principle widely discussed in the SEM literature (e.g., Raykov & Marcoulides, 1999, and references therein). Hence, Popperian logic, which maintains that a goal of empirical science is to formulate theories that can be falsified, is facilitated by an application of the parsimony principle. If a more parsimonious model is found to be acceptable, then one may also place more trust in it because it has withstood a higher chance of rejection than a less parsimonious model. Of course, researchers are cautioned that rigid and routine applications of the parsimony principle can lead to conclusions favoring an incorrect model and implications that are incompatible with those of the correct model (for further discussion see Raykov & Marcoulides, 1999).

The chi-square value T has received a lengthy discussion in this section for two reasons. First, historically and traditionally, it has been the index that has attracted the most attention over the past 30 years or so. In fact, most of the fit indices devised more recently in the SEM literature are in some form functions of the chi-square value. Second, the chi-square value has the important feature of being an inferential fit index. That is to say, by using it one is in a position to make a generalization about the fit of the model in the studied population. The reason for this is that the large-sample distribution of T in the population is known (namely central chi-square, with a correct model fitted to the covariance matrix), and a p value can be attached to each particular sample's value of T. This feature is not shared by many of the other fit indices used in SEM.

However, this statistical evaluation of the plausibility of a model using the chi-square value T cannot always be strictly followed. This is due to the fact that with very large samples T cannot be really relied on. The reason is readily seen from its definition in Equation 7. Because the value of T is obtained by multiplying $N - 1$ (the sample size less 1) by the minimum of the fit function, increasing the sample size generally leads to an increase in T as well. And yet, the model's degrees of freedom remain the same (because the proposed model hasn't changed) and, hence, so does the reference chi-square distribution against which T is judged for significance. Consequently, with very large samples there is a spurious tendency to obtain large values of T, which tend to be associated with small p values. Hence, if one were to use only the chi-square value's p value as an index of model fit, there will be an artificial tendency with very large samples to reject the model even if it were only marginally inconsistent with the data.

Alternatively, there is another spurious tendency with very small samples for T to remain small and to be associated with larger p values (suggesting the model as a plausible data-description means). Thus, the chi-square value and its p value alone cannot be fully trusted in the general case of model evaluation. Other fit indices must also be examined in order to obtain a better picture of model fit.

Descriptive-Fit Indices. The limitations of the chi-square value indicate the importance of alternative fit indices to aid in the process of model evaluation. The descriptive-fit indices provide an alternative family of indices that assess the goodness-of-fit of the proposed model based on the particular sample at hand.

The first descriptive-fit index ever proposed is called the goodness-of-fit index (GFI). This index can be loosely considered to be a measure of the proportion of variance and covariance that the proposed model is able to explain (similar to R^2 in a regression analysis). If the number of parameters is also taken into account in computing this measure, the resulting index is called the adjusted goodness-of-fit index (AGFI) (similar to the adjusted R^2 used in a regression analysis). The GFI and the AGFI descriptive indices range between 0 and 1, and are usually fairly close to 1 for well-fitting models. Unfortunately, as with many other descriptive indices currently used, there are no strict norms for the GFI and AGFI below which a model cannot be considered a plausible description of the analyzed data and above which one can rest assured that the model approximates the data reasonably well. As a rough guide, it is currently viewed that models with a GFI and AGFI in the mid-.90s or above may well represent a reasonably good approximation of the data (Hu & Bentler, 1999).

There are two other descriptive indices that are also very useful for model-fit evaluation. These are the normed fit index (NFI) and the non-normed fit index (NNFI) (Bentler & Bonnet, 1980). The NFI and NNFI are based on the idea of comparing the proposed model to a model in which absolutely no interrelationships are assumed among any of the variables. This model with no relationships is referred to as the independence model or the null model. The name independence or null derives from the fact that the model assumes the variables only have some variances, but that there are no relationships among them. Thus, the null model represents the extreme case of no relationships and the interest lies in comparing the proposed model to the null model. If the chi-square value of the null model is compared to that of the proposed model, one should get an idea of how much better the proposed model fits the data relative to how bad it could possibly be (i.e., relative to the null model). This is the basic idea that underlies the NFI and NNFI descriptive-fit indices.

The NFI is computed by relating the difference of the chi-square value for the proposed model to the chi-square value for the independence or null model. The NNFI is a simple variant of the NFI that takes into account the degrees of freedom of the proposed model. This is done in order to take into account model complexity, as reflected in the degrees of freedom of the proposed model. This is because more complex models have more parameters and hence fewer degrees of freedom, whereas less complex models have less parameters and thus more degrees of freedom. For this reason, one can consider the degrees of freedom as indicators of the complexity of a model.

Similar to the GFI and AGFI, models with NFI and NNFI close to 1 are considered to be more plausible means of describing the data than models for which these indices are further away from 1. Unfortunately, once again, there are no strict norms above which one can consider the indices as supporting model plausibility or below which one can safely reject the model as a means of data description. As a rough guide, models with NNFI and NFI in the mid-.90s or higher are viewed likely to represent a reasonably good approximation of the data (Hu & Bentler, 1999).

In addition to the GFI, AGFI, NNFI, and NFI, there are more than a dozen other descriptive-fit indices that have been proposed in the SEM literature over the past 20 years or so. Despite this plethora of descriptive-fit indices, it turns out that most of them are directly related to the chi-square value T and simply represent reexpressions of it or its relationships to other models' chi-square values and related quantities. The interested reader may refer to more advanced SEM books that provide mathematical definitions of each (e.g., Bollen, 1989; Marcoulides & Hershberger, 1997; Marcoulides & Schumacker, 1996), as well as the program manuals for EQS and LISREL (Bentler, 1995; Jöreskog & Sörbom, 1993a).

Alternative-Fit Indices.   Alternative-fit indices are based on an altogether different conceptual approach to the process of hypothesis testing in SEM, which is referred to as an alternative approach to model assessment. Alternative-fit indices have been developed over the past 20 years and largely originate from an insightful paper by Steiger and Lind (1980). The basis for alternative-fit indices is the noncentrality parameter (NCP), denoted δ. The NCP basically reflects the extent to which a proposed model does not fit the data. For example, if a proposed model is correct and the sample is large, the test statistic T presented in Equation 7 follows a (central) chi-square distribution, but if the model is not correct (i.e., is misspecified) then T follows a noncentral chi-square distribution. As an approximation, a noncentral chi-square distribution can be thought of as resulting when the central chi-square distribution is shifted to the right by δ units. In this way, the NCP can be viewed as an index reflecting the degree to which the model fails to fit the data. Thus, the larger the NCP, the worse the model and the smaller the NCP, the better the model. It can be shown that with not-too-misspecified models, normal data, and large samples, δ approximately equals $(N-1)F_{ML,0}$, where $F_{ML,0}$ is the value of the fit function based on the maximum likelihood method of estimation. In practice, δ is estimated by $\hat{\delta} = (T-d)/n$, where d is equal to the model degrees of freedom, and $n = N-1$ is the sample size less 1, or is estimated by 0 if $T < d$.

Within the alternative approach to model testing, the conventional null hypothesis that a proposed model perfectly fits the covariance matrix is relaxed. This is explained by the observation that in practice every model is wrong even before it is fitted to the data. Indeed, the reason why a model is used when studying a phenomenon of interest is that the model should represent a useful simplification and approximation of reality rather then a precise replica of it. That is, by its very nature, a model cannot be correct because then it would have to be an exact copy of reality and therefore be useless. As such, in the alternative approach to model testing, the conventional null hypothesis of perfect model fit (traditionally tested in SEM by examining the chi-square index and its p value) is really of no interest. Instead, one is primarily concerned with evaluating the extent to which the model fails to fit the data. Consequently, for the reasonableness of a model as a means of data description, one should impose weaker requirements for degree of fit.

This is the logic of model testing that is followed by the root mean square error of approximation (RMSEA) index that has recently become quite a popular index of model fit. The RMSEA is defined as

$$\pi = \sqrt{\frac{(T-d)}{(dn)}} \qquad (8)$$

when $T \geq d$, or as 0 if $T < d$. The RMSEA, similar to other fit indices, also takes into account model complexity, as reflected in the degrees of freedom. Some researchers have suggested that a value of the RMSEA of less than .05 is indicative of the model being a reasonable approximation to the data (Browne & Cudeck, 1993). Some researchers have also argued that, because in Equation 8 a division by sample size occurs, the RMSEA is not sample-dependent (unlike the chi-square value). In fact, it is believed that this feature of the RMSEA sets it apart from many other fit indices that are sample-dependent or have characteristics of their distribution (such as the mean) that depend on sample size (Marsh et al., 1996).

The RMSEA is not the only index that can be obtained as a direct function of the noncentrality parameter. The comparative-fit index (CFI) also follows the logic of comparing a proposed model with the null model assuming no relationships between the measures (Bentler, 1990). The CFI is defined as the ratio of improvement in noncentrality (moving from the null to the proposed model) to the noncentrality of the null model. Typically, the null model has considerably higher noncentrality than a proposed model because it is expected to fit the data poorly. Thus, values of CFI close to 1 are considered likely to be indicative of a reasonably well-fitting model. Again, there are no norms about how high the CFI should be in order to safely retain or reject a proposed model. In general, a CFI in the mid-.90s or above is usually associated with models that are plausible approximations of the data.

The expected cross-validation index (ECVI) was also introduced as a function of the noncentrality parameter (Browne & Cudeck, 1993). It represents a measure of the degree to which one would expect a proposed model to replicate in another sample from the same population. In a set of several proposed models of the same phenomenon, a model is preferred if it minimizes the value of ECVI relative to other models. The ECVI was developed partly as a reaction to the fact that because the RMSEA presumably does not depend on sample size, it cannot account for the fact that with small samples it would be unwise to fit a very complex model (i.e., with many parameters). The ECVI accounts for this possibility, and when the maximum likelihood method of estimation is used it will be closely related to the Akaike information criterion (AIC). The AIC is a special kind of fit index that takes into account both the measure of fit and model complexity (Akaike, 1987). Generally, proposed models with lower values of ECVI and AIC are more likely to be better means of data description than models with higher ECVI and AIC indices. The ECVI and AIC have become quite popular in SEM applications, particularly for the purposes of examining competing models (i.e., when a researcher is considering several models and wishes to select from them the one with the best fit).

Another important feature of this alternative approach to model assessment involves the routine use of confidence intervals. Recall from basic statistics that a confidence interval provides a range of plausible values for the population parameter being estimated at a given confidence level. The width of the interval is also indicative of the precision of estimation of the parameter. Of special interest to the alternative approach to model testing is the left endpoint of the 90% confidence interval of the RMSEA index. Specifically, if this limit of the RMSEA is considerably smaller than .05 and the interval not too wide, it can be argued that the model is a plausible means of describing the analyzed data. Thus, if the RMSEA is smaller than .05 or the left endpoint of its confidence interval markedly smaller than .05 (with this interval not excessively wide), the model can be considered a reasonable approximation of the data.

In concluding this section on model testing and fit evaluation, it must be emphasized that no decision should be based on a single index, no matter how favorable for the model the index may appear. As indicated earlier, every index represents a certain aspect of the fit of a proposed model, and in this sense is a source of limited information as to how good the model is or how well it can be expected to perform in the future. Therefore, a decision to reject or retain a model should always be based on multiple goodness-of-fit indices (and if possible on the results of replication studies). In addition, as indicated in the next section, important insights regarding model fit can sometimes be obtained by also conducting an analysis of residuals.

Analysis of Residuals

All fit indices considered in the previous section should be considered overall measures of model fit. In other words, they are all summary measures of fit and none of them provide information about the fit of individual parts of the model. As a consequence, it is possible for a proposed model to be seriously misspecified in some parts (i.e., incorrect with regard to some of the variables and their relationships), but be very well fitting in other parts, and for an evaluation of the previously discussed fit criteria to suggest that the model is plausible.

For example, consider a model that is substantially off the mark with respect to including the presence of a relationship between two particular variables (i.e., the model does actually exclude this relationship). In such a case, the difference between the sample covariance and the covariance reproduced by the model at the final solution—called residual for that pair of variables—may be substantial. This result would suggest that the model cannot be considered a plausible means of data description. However, at the same time, the model may do an excellent job of explaining all of the remaining covariances and variances in the sample covariance matrix S and result in a nonsignificant chi-square value and favorable other fit indices. Such an apparent paradox may emerge because the chi-square value T (or any of the other fit indices discussed) is a measure of overall fit. Thus, all that is provided by overall measures of model fit is a summary picture of how well the proposed model fits the whole analyzed matrix, but no information is provided about how well the model reproduces the individual elements of that matrix.

To counteract this possibility, two types of residuals can be examined in most SEM models. The unstandardized residuals index the amount of unexplained covariance and variance in terms of the original metric of the raw data. However, if the original metric of the raw data is quite different across measured variables, it is impossible to examine residuals and determine which are large and which are small. A standardization of the residuals to a common metric, as reflected in the standardized residuals, makes the comparison much easier.

A standardized residual close to or above +3 indicates that the model considerably underexplains a particular relationship between two variables. Conversely, a standardized residual close to or below –3 indicates that the model markedly overexplains the relationship between the two variables. Using this residual information, a researcher may decide to either add or remove some substantively meaningful paths or covariances, which could contribute to a smaller residual associated with the involved two variables and hence a better-fitting model with regard to their relationship.

Overall, good-fitting models will typically exhibit a stem-and-leaf plot of standardized residuals that closely resembles a symmetric distribution. In addition, examining a Q plot of the standardized residuals is a useful means of checking the plausibility of a proposed model. The Q plot graphs the standardized residuals against their expectations should the model be a good means of data description. With good-fitting models, a dotted line through the marks of the residuals on that plot will be close to vertical (Jöreskog & Sörbom, 1993c). Any serious departures from a straight line indicate either serious model misspecifications or violations of the normality assumption (e.g., nonlinear trends in the relationships between some observed variables; see Raykov, in press).

An important current limitation in SEM is the infrequent evaluation of estimated individual-case residuals. Individual-case residuals are routinely used in applications of regression analysis because they help researchers with model evaluation and modification. In regression analysis, residuals are defined as the differences between individual raw data and their model-based predictions. Unfortunately, SEM developers have only recently begun to investigate more formally ways in which individual-case residuals can be defined within this framework. The development of ways for defining individual-case residuals is also hampered by the fact that most structural models are based on latent variables, which cannot be directly observed or precisely measured. Therefore, very important pieces of information that are needed in order to arrive at individual-case residuals similar to those used in regression analysis are often missing.

Modification Indices

A researcher usually conducts a SEM analysis by fitting a proposed model to the available data. If the specified model does not fit, one may accept this fact and leave it at that or may consider the question, "How should the specified model be modified to improve the fit?" In the SEM literature, the modification of a specified model in order to improve fit has been termed a specification search (Long, 1983). Accordingly, a specification search is conducted with the intent to detect and correct the specification error between a proposed model and the true model characterizing the population and variables in the study. Although in theory researchers should fully specify and deductively hypothesize a model prior to data collection and model testing, in practice this is often simply not possible, either because a theory is poorly formulated or because it is altogether nonexistent. As a consequence, specification searches have become a common practice in SEM applications. In fact, most currently available SEM program provide researchers with options to conduct specification searches to improve model fit, and some new search procedures have also been developed to automate this process (see Marcoulides, Drezner, & Schumacker, 1998; Scheines, Spirtes, & Glymour, in press; Scheines, Spirtes, Glymour, Meek, & Richardson, 1998; Spirtes, Scheines, & Glymour, 1990).

Specification searches are clearly helpful for improving a model that is not fundamentally misspecified, but is incorrect only to the extent that it has some missing paths or some of its parameters are involved in unnecessarily restrictive constraints. With such models, it can be hypothesized that the unsatisfactory fit stems from overly strong restriction(s) on its parameters, that they are either fixed to 0 or set equal to other parameter(s) (or included in a more complex relationship). Of course, the application of any means of model improvement is only appropriate when the model modification suggested is theoretically sound and does not contradict the results of previous research in the particular substantive domain. Alternatively, the results of any specification search that do not agree with past research should be subjected to an analysis based on new data before any real validity can be claimed.

The indices that can be used as diagnostic statistics about which parameters should be changed in a model are called modification indices (a term used in the LISREL program) or Lagrange multiplier statistics (a term used in the EQS program). The value of a *modification index* (the term is used generically) indicates how much a proposed model's chi-square would decrease if a particular parameter were freely estimated or freed from a constraint in the preceding model run. There is also another modification index, called the Wald index, which takes a slightly different approach to the problem. The value of the *Wald index* indicates how much a proposed model's chi-square would increase if a particular parameter were fixed to 0 (i.e., if the parameter were deleted from the proposed model).

The modification indices address the question of how to improve an initially specified model that does not fit satisfactorily the data. Although no strict rules-of-thumb exist concerning how large the indices must be to warrant a meaningful model modification, based on purely statistical considerations one might simply consider making changes to parameters associated with the highest modification indices. If there are several parameters with high modification indices, one should consider freeing them one at a time, beginning with the largest, because it is well known that a single change in a model can affect other parts of the solution (Jöreskog & Sörbom, 1990; Marcoulides et al., 1998). When LISREL is used, modification indices larger than 5 generally merit close consideration. Similarly, when EQS is used, parameters associated with significant Lagrange-multiplier or Wald-index statistics also deserve close consideration.

It must be emphasized, however, that any model modification first must be justified on theoretical grounds and be consistent with already available theories or results from previous research in the substantive domain, and only second must be in agreement with statistical optimality criteria such as those mentioned. Blind use of modification indices can turn out to be a road to such models that lead researchers astray from their original substantive goals. It is therefore imperative to consider changing only those parameters that have a clear substantive interpretation. As discussed in later chapters, additional statistics, in the form of the estimated change for each parameter, can also be taken into account before one reaches a final decision regarding model modification.

In conclusion, it is important to emphasize that results obtained from any model-improvement specification search may be unique to the particular data set and that capitalization on chance can occur during the search. Consequently, once a specification search is conducted, a researcher is entering a more exploratory phase of analysis. This has also purely statistical implications in terms of not keeping the overall significance level at the initially prescribed nominal value (the preset significance, usually .05). Hence, the possibility exists of arriving at such statistically significant results regarding aspects of the model due only to chance fluctuations. Thus, the likelihood of falsely declaring at least one of the conducted statistical tests of the model to be significant is increased rather than being the same as that of any single test. For this reason, any models that result from specification searches must be cross-validated before any real validity can be claimed.

CHAPTER TWO

Getting to Know the EQS and LISREL Programs

In chapter 1, the basic concepts of structural equation modeling methodology were introduced. In this chapter, the essential elements of the notation and syntax used in the EQS and LISREL programs are introduced. The chapter begins by presenting an easy-to-follow flow chart that depicts the general principle behind constructing input files for each program (although this flow chart can actually be applied with any SEM program language). This material is followed by a discussion of the most important elements of the EQS and LISREL program languages. The EQS program is presented first because the preceding chapter has already familiarized the reader with a number of important concepts that substantially facilitate an introduction to it. The LISREL program, with its slightly different structure, is dealt with second. For more extensive discussions and information, the reader is referred to the latest versions of the program manuals (Bentler, 2000; Jöreskog & Sörbom, 1999).

STRUCTURE OF INPUT FILES FOR SEM PROGRAMS

Each SEM program can essentially be considered a new language with special syntax and rules that must be precisely followed. The special syntax is used to tell the program all the details about the observed data and the proposed models. The details about the data and models are communicated to the program as input commands. The results of the program acting on these commands are then presented in the output.

Obviously, the most essential piece of information that must be provided to the program is how the proposed model looks. Conveying infor-

mation about the model begins with a listing of each observed and unobserved variable as well as the way they relate to one another. It is just as
important that the program be informed about the parameters of the
model. As discussed in the previous chapter, each program has in its memory the formal way the model reproduced matrix Σ can be obtained, once
it is informed about the model and its parameters. Thus, as illustrated in
chapter 1, one must first draw the path diagram of the model and then define all the necessary parameters using Rules 1 to 6 before proceeding
with the computer analysis.

In addition to this, one must also communicate to the program a few
details about the data on which the proposed model will be fit. In particular, information concerning the location of the data, the name of the file,
and the format of the data must be provided. For example, with respect to
the format of the data, it must be specified whether the data is in raw form,
in the form of a covariance matrix (and means, if needed; see chap. 6), or
in the form of a correlation matrix (and variable standard deviations). It is
also important to explicitly state the number of variables in the data set
and provide information about the observed sample size. Fortunately,
most SEM programs will echo in their output the structure of the input file
(in terms of the number of variables, observations, etc.) and also display
the data to be analyzed in the form of either a covariance or a correlation
matrix. Thus, a user can quickly check the output to ensure that the program has indeed correctly read the data to be analyzed.

Finally, one must communicate to a SEM program which type of analysis is desired. For example, particulars about the estimation-fitting process
(e.g., maximum likelihood or weighted least squares), the number of iterations to be conducted, or the specific measures of model fit to be considered must be provided. Although many of the details can be implemented
automatically or handled by using a program's default settings, in order to
have some sense of control over a program, it is always better for a user to
specify the particulars of a proposed model.

In simple terms, the flow chart here represents the backbone of any input file of a model to be fitted with a SEM program.

Location and form of data to be analyzed

↓

Description of proposed model to be fitted

↓

User-specified information about the final solution

The next two sections follow this flow chart for setting up input files in
EQS and LISREL. Although this is done first using only the most necessary
information, it will prove sufficient for fitting most of the models used in
the book. The discussion is extended when more complicated models are

discussed in later chapters (such as when dealing with multisample or mean structure analyses).

INTRODUCTION TO THE EQS NOTATION AND SYNTAX

The discussion in chapter 1 has already laid the foundation for introducing the specific elements needed to set up input files using the EQS syntax and notation. An input file in EQS is made up of various command lines. The beginning of each command line is signaled by a forward slash (/), and its end is signaled by a semicolon (;). To begin the introduction to EQS, a list of the individual commands that are most often used in practice when conducting SEM analyses are presented next (for further details see Bentler, 1995 or 2000). Each command is illustrated using the data and proposed factor analysis model originally displayed in Fig. 6 in chapter 1. For ease of presentation, the same figure is displayed again in Fig. 7. In addition, and to keep all program input files visually separate from the regular text, all command lines are capitalized throughout the book.

Title Command

One of the first things needed to create an EQS input file is a title command line. This line simply describes in plain English the type of model examined (e.g., a confirmatory factor analysis model or a path analytic model) and perhaps some of its specifics (e.g., a study of college students or middle managers). The title command line is created by the keyword /TITLE. On the line immediately following, and for as many lines as needed, an explanatory title is provided. For example, suppose that the model being studied is the factor analysis model displayed in Fig. 7. The title command line could then be listed as

/TITLE
FACTOR ANALYSIS MODEL OF THREE INTERRELATED CONSTRUCTS
EACH MEASURED BY THREE INDICATORS;

It is important to emphasize that although the description in a title command line can be kept to a single line, the more details that are provided the better one will recall the purpose of the study, especially when revisiting input files months (or even years) later.

Data Command

The data command line provides details about the data to be analyzed. This is done by using the keyword /SPECIFICATIONS. On the next line, the exact number of variables in the data set are provided using the key-

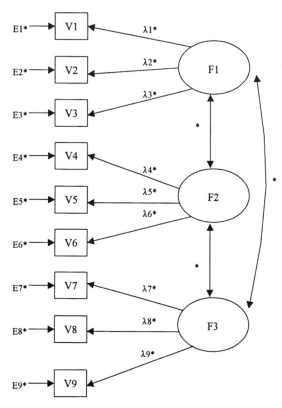

FIG. 7. Example factor analysis model. F_1 = Parental dominance; F_2 = Child intelligence; F_3 = Achievement motivation.

word VARIABLES= , followed by the number. Then information about the number of observations in the data set is provided using the keyword CASES= , followed by the sample size. The data to be used in the study may be placed directly in the input file or in a separate data file. If the data are available as a covariance matrix, the keyword MATRIX=COV can be used (although it is the default option in EQS). If the data to be examined are in raw form (e.g., one column per variable), then the keyword MATRIX=RAW is used and the name of the file enclosed in apostrophes (and its location) are provided with the keyword DATA_FILE= (e.g., DATA_FILE='C:/DATA/DATAFILE'). If the data to be examined in the form of a covariance matrix are placed directly in the input file (e.g., the covariance matrix S presented in chap. 1), the keyword /MATRIX is used followed by the matrix itself. Although the covariance matrix will eventually appear at the very end of the input file (see final program), for clarity it is discussed now in the beginning part of the file. If a matrix of variable interrelationships other than the covariance matrix is to be analyzed, this information can be provided using the keyword ANALYSIS= , followed by MOMENTS if the covariance or mean structure is analyzed (i.e., the means of the variables along with their sample covariance matrix S), or by COR-RELATION if the correlation matrix is analyzed (in EQS6).

The default method of estimation in EQS is maximum likelihood (ML). If a method other than ML is to be used for estimation, it is stated using the keyword METHOD= , followed by the abbreviation of the method recognizable by the program (e.g., GLS or LS). Although EQS provides the option of selecting from among several estimation procedures, as indicated in chapter 1, only output from the maximum likelihood method will be provided in this book. Using the proposed factor analysis model in Fig. 7 as an illustration, the data command line can then be listed as

```
/SPECIFICATIONS
VARIABLES=9;
CASES=245;
METHOD=ML;
MATRIX=COV;
ANALYSIS=COV;
/MATRIX
1.01
.32   1.50
.43   .40   1.22
.38   .25   .33   1.13
.30   .20   .30   .7    1.06
.33   .22   .38   .72   .69   1.12
.20   .08   .07   .20   .27   .20   1.30
.33   .19   .22   .09   .22   .12   .69   1.07
.52   .27   .36   .33   .37   .29   .50   .62   1.16
```

It should be noted that because each of the command lines ends in a semicolon, they can also be specified on a single line. The only command line in which no semicolon is needed to mark the end of the line is the /MATRIX command.

Model-Definition Commands

The next command lines needed in the EQS input file (following the flow chart presented earlier) have to do with model description. To accomplish this one must take a close look at the path diagram of the proposed model and provide the information about the specified parameters in the model. In summary, setting up the model definition commands involves the following three activities: (a) writing out the equations relating each dependent variable to its explanatory variables, (b) determining the status of all variances of independent variables (whether free or fixed, and at what values), and (c) determining the status of the covariances for all independent variables. These three activities are accomplished by using the keywords /EQUATIONS, /VARIANCES, and /COVARIANCES.

Each free parameter in the model is denoted in the program language by an asterisk (any not mentioned are automatically assumed to be 0). Following closely the model path diagram in Fig. 7 should result in 21 asterisks appearing in the model-definition command lines. Thus, the /EQUATIONS command is

```
/EQUATIONS
V1 = *F1 + E1;
V2 = *F1 + E2;
V3 = *F1 + E3;
V4 = *F2 + E4;
V5 = *F2 + E5;
V6 = *F2 + E6;
V7 = *F3 + E7;
V8 = *F3 + E8;
V9 = *F3 + E9;
```

It is important to note that the model parameters (i.e., the nine λ's in Equation 1 in chap. 1) have simply been changed to asterisks. In addition, if needed, one can also assign a special start value to any of the model parameters by writing that value immediately before the asterisk in the equation (e.g., V9 = .9*F3 + E9).

The keyword /VARIANCES is used to inform the program about the status of the 12 independent variable variances (recall Rule 1 in chap. 1). According to Fig. 7, there are nine residual variances and three factor variances (which, following Rule 6, will be fixed to 1 in order to insure that their metric is set). Thus, the following command line is used

```
/VARIANCES
F1 TO F3 = 1; E1 TO E9 = *;
```

Note that one can use the TO convention in order to save tedious writing of all independent variables in the model.

Finally, information about the three factor covariances (correlations) included in the model must be communicated. This is provided as

```
/COVARIANCES
F2,F1=*; F3,F1=*; F3,F2=*;
```

Once the model-definition command lines have been completed, it is important to ensure that Rules 5 and 6 have not been contradicted in the input file. Thus, a final check ensures that each of the three factor variances is fixed at 1 (Rule 6), and that no variance or covariance of a dependent variable and no covariance of a dependent and an independent variable have been declared model parameters (Rule 5). Counting the number of asterisks in the model-definition command lines, one finds 21 model parameters appearing in the input file—just as many as in the path diagram of the model in Fig. 7. Obviously, if the two numbers differ, some parameter has either been left out or incorrectly declared.

Complete EQS Input File for the Model in Fig. 7

Based on the command lines, the following complete EQS input file emerges. The use of the keyword /END signals the end of the EQS input file.

```
/TITLE
FACTOR ANALYSIS MODEL OF THREE INTERRELATED CONSTRUCTS
EACH MEASURED BY THREE INDICATORS;
/SPECIFICATIONS
VARIABLES=9; CASES=245; METHOD=ML; MATRIX=COV;
ANALYSIS=COV;
/EQUATIONS
V1 = *F1 + E1;
V2 = *F1 + E2;
V3 = *F1 + E3;
V4 = *F2 + E4;
V5 = *F2 + E5;
V6 = *F2 + E6;
V7 = *F3 + E7;
V8 = *F3 + E8;
V9 = *F3 + E9;
/VARIANCES
F1 TO F3 = 1; E1 TO E9 = *;
/COVARIANCES
F1,F2=*; F1,F3=*; F2,F3=*;
/MATRIX
1.01
.32   1.50
.43   .40   1.22
.38   .25   .33   1.13
.30   .20   .30   .7    1.06
.33   .22   .38   .72   .69   1.12
.20   .08   .07   .20   .27   .20   1.30
.33   .19   .22   .09   .22   .12   .69   1.07
.52   .27   .36   .33   .37   .29   .50   .62   1.16
/END;
```

A Useful Abbreviation

Each of the command lines in this EQS input file can also be abbreviated, usually to the first three letters of the keyword. This often saves a considerable amount of time for the researcher setting up numerous input files. Therefore, the following EQS input file can also be used:

```
/TIT
FACTOR ANALYSIS MODEL OF THREE INTERRELATED CONSTRUCTS
EACH MEASURED BY THREE INDICATORS;
/SPE
VAR=9; CAS=245; MET=ML; MAT=COV; ANA=COV;
/EQU
V1 = *F1 + E1;
V2 = *F1 + E2;
V3 = *F1 + E3;
V4 = *F2 + E4;
V5 = *F2 + E5;
V6 = *F2 + E6;
V7 = *F3 + E7;
V8 = *F3 + E8;
V9 = *F3 + E9;
/VAR
F1 TO F3 = 1; E1 TO E9 = *;
/COV
F1,F2=*; F1,F3=*; F2,F3=*;
/MAT
1.01
.32   1.50
.43   .40   1.22
.38   .25   .33   1.13
.30   .20   .30   .7    1.06
.33   .22   .38   .72   .69   1.12
.20   .08   .07   .20   .27   .20   1.30
.33   .19   .22   .09   .22   .12   .69   1.07
.52   .27   .36   .33   .37   .29   .50   .62   1.16
/END;
```

Imposing Parameter Restrictions

The example here does not include any user-specified information about generating a final solution (i.e., the last part of the simple flow chart, regarding special output) because no output information beyond that provided by EQS by default was needed. However, later in this book such specifications, which either relate to the execution of the iteration process or ask the program to list information that it otherwise does not provide automatically, are included. For example, suppose that one is interested in testing the plausibility of the hypothesis that the first three observed variables—the indicators of the Parental domination factor—all have equal factor loadings (in the psychometric literature, this would be referred to as a triplet of tau-equivalent tests). Such an assumption is tantamount to the three measures assessing the construct in the same units of measurement. In order to introduce this constraint to the EQS program, a new command-line section dealing specifically with parameter restrictions is included in the input file. The command line uses the keyword /CONSTRAINTS and contains the following specification of the imposed parameter equalities:

/CONSTRAINTS
(V1,F1) = (V2,F1) = (V3,F1);

For consistency, it is suggested that all constraints imposed on the model be included immediately after the /COVARIANCE command line, which usually ends the model-definition part of an EQS input file.

INTRODUCTION TO THE LISREL
NOTATION AND SYNTAX

This section deals with the notation and syntax used in the general LISREL model (the submodel 3B in the LISREL manual; Jöreskog & Sörbom, 1993b). In order to keep things simple, the same factor analysis model examined in the previous section is considered when introducing the various LISREL command lines. Although the particular notation and syntax of the LISREL command lines are quite different to those of EQS, the specific elements needed to set up the input files are very similar.

The general LISREL model assumes that a set of observed variables (denoted as Y) is used to measure a set of latent variables (denoted as η—the lowercase Greek letter *eta*). The relationships between the observed and latent variables are represented by a factor analysis model (also referred to as a measurement model), whose residual terms are denoted by ε (the lowercase Greek letter *epsilon*).[1] The explanatory relationships among the

---

[1] In LISREL, the observed variables can actually be denoted either by Y or X, the latent variables either by η or ξ (the lowercase Greek letter *eta* or *ksi*), and the residual terms either by ε or δ (the lowercase Greek letter *epsilon* or *delta*). For ease of presentation, however, only Y, η, and ε are used here—the notation used in the submodel 3B in the LISREL manual. Of course, as one becomes more familiar with the LISREL program, selecting which notation to use for representing a measurement SEM model will become simply a matter of taste. Either one can be used interchangeably. However, when the SEM model includes both a measurement part and a structural part, X is usually used to denote the observed variables of the predictor latent variables and Y is used for the observed variables of the outcome latent variables.

latent variables constitute the structural model. To avoid any confusion, however, throughout this book these two models are referred to as the *measurement* and *structural parts* of a structural equation model.

Measurement Part Notation

Consider the factor analysis example displayed in Fig. 7. The measurement part of this factor analysis model can be written using the following notation (note the slight deviation in notation from Equation 1 in chap. 1):

$$
\begin{aligned}
Y_1 &= \lambda_{11}\eta_1 + \varepsilon_1, \\
Y_2 &= \lambda_{21}\eta_1 + \varepsilon_2, \\
Y_3 &= \lambda_{31}\eta_1 + \varepsilon_3, \\
Y_4 &= \lambda_{42}\eta_2 + \varepsilon_4, \\
Y_5 &= \lambda_{52}\eta_2 + \varepsilon_5, \\
Y_6 &= \lambda_{62}\eta_2 + \varepsilon_6, \\
Y_7 &= \lambda_{73}\eta_3 + \varepsilon_7, \\
Y_8 &= \lambda_{83}\eta_3 + \varepsilon_8, \\
Y_9 &= \lambda_{93}\eta_3 + \varepsilon_9.
\end{aligned}
\tag{9}
$$

To facilitate the discussion, the model displayed in Fig. 7 is reproduced again in Fig. 8 using the new notation. It is important to note that this is the same model, only the notation differs.

Note that Equations 9 are formally derived from Equations 1 presented in chapter 1 after several simple modifications. First, change the symbols of the observed variables from V_1 through V_9 to Y_1 through Y_9, respectively. Then change the symbols of the factors from F_1 through F_3 to η_1 through η_3, respectively. Next, change the symbols of the residual terms from E_1 through E_9 to ε_1 through ε_9, respectively. Finally, add a second subscript to the factor loadings, which is identical to the factor on which the manifest variable loads.

The last step (changing to the double-index notation) represents a rather useful notation used in developing LISREL input files. Each factor loading or regression coefficient in a proposed model is numbered using two subscripts. The first corresponds to the number (index) of the dependent variable, and the second corresponds to the number of the independent variable in the pertinent equation. For example, in the fifth equation of Equations 9, $Y_5 = \lambda_{52}\eta_2 + \varepsilon_5$, the factor loading λ has two subscripts; the first corresponds to the dependent variable (Y_5), and second is the index of the latent variable (η_2). Independent variable variances and covariances have as subscripts the indices of the variables they are related to. Therefore, every variance has the index of the variable it belongs to as

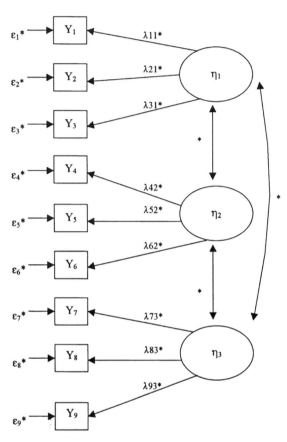

FIG. 8.   Example factor analysis model using LISREL notation. η_1 = Parental dominance; η_2 = Child intelligence; η_3 = Achievement motivation.

subscripts twice, whereas a covariance has the indices of the variables it relates as subscripts.

Structural Part Notation

In the example model displayed in Fig. 8, there is no structural part because no explanatory relationships are assumed among any of the constructs. If one assumed, however, that the latent variable η_2 was regressed on η_1 and that η_3 was regressed on η_2, then the structural part for the model would be

$$\eta_2 = \beta_{21}\eta_1 + \zeta_2,$$
$$\eta_3 = \beta_{32}\eta_2 + \zeta_3. \tag{10}$$

In Equation 10 the structural slopes of the two regressions are denoted by β (the Greek letter *beta*), whereas the corresponding residual terms are symbolized by ζ (the Greek letter *zeta*). Note again the double indexing of the β, in which, as mentioned earlier, the first index denotes the dependent variable and the second denotes the independent variables in the pertinent equation. The indices of the ζ are identical to those of the latent dependent variables to which they belong as residual terms.

Two-Letter Abbreviations

A characteristic feature of the LISREL notation and syntax is the use of abbreviations based on the first two letters of keywords or parameter names. For example, within the general LISREL model, a factor loading can be referred to using the notation LY, followed by its indices. Similarly, a structural regression slope is referred to using BE (for *beta*), followed by its indices. Thus, using the indexing principle, the loading of the fifth manifest variable on the second factor is denoted by LY(5,2). Similarly, the slope of the third factor when regressed on the second factor is denoted by BE(3,2) (although the brackets and delimiting comma are not required, they make the presentation easier to follow).

Variances and covariances between (independent) latent variables are denoted in the general LISREL model by PS (the Greek letter *psi*, ψ), followed by the indices of the constructs involved. For example, the covariance between the first two factors in Fig. 8 is denoted by PS(2,1), whereas the variance of the third factor is PS(3,3). Finally, variances and covariances for residual terms are denoted by TE (the Greek letters *theta* and *epsilon*, θ ε). For instance, the variance of the seventh residual term ε_7 is symbolized by TE(7,7), whereas the covariance (assuming such a covariance exists) between the first and fourth error terms would be denoted by TE(4,1).

Thus, the general LISREL model parameters needed to examine a structural equation model are (a) factor loadings as LY's, (b) structural regression coefficients as BE's, (c) latent-variable variances and covariances as PS's, and (d) residual variances and covariances as TE's.

Matrices of Parameters—A Helpful Notation Tool

The representation of the indices of model parameters as numbers delineated by a comma and placed in brackets strongly resembles that of matrix elements. Indeed, in LISREL notation, parameters can be thought of as being conveniently collected into matrices. For example, the factor loading λ_{42}, denoted in LISREL notation as LY(4,2), can be thought of as the second element (from left to right) in the fourth row (from top to bottom) of the matrix denoted LY. Similarly, the structural regression slope of η_3 on

η_1, β_{31}, denoted in LISREL notation as BE(3,1), is the first element in the third row of the matrix BE. The covariance between the third and fifth factors, ψ_{53}, is the third element in the fifth row of the matrix PS, PS(5,3), whereas θ_{21} is the first element in the second row of the matrix TE, TE(2,1) or simply TE 2 1. The matrices can be rectangular or square, be symmetric or nonsymmetric (commonly referred to as full matrices), have only 0 elements, or be diagonal.

In the general LISREL model notation, all factor loadings are considered elements of a matrix called LY, which is usually a rectangle (rather than square) matrix because in practice there are frequently more observed variables than factors. Furthermore, the structural regression coefficients (the regression coefficients when predicting a latent variable in terms of other ones) are elements of a matrix called BE. The BE matrix is square because it represents the explanatory relationships among the set of latent variables. Finally, the factor variances and covariances are entries of a (symmetric) square matrix PS, whereas the error variances and covariances are the elements of a (symmetric) square matrix TE. Of course, unless one has strong theoretical justification for considering the presence of covariances among errors, the matrix is usually assumed to be diagonal (i.e., to only include the error variances of the observed variables in the model). Thus, in order to use the general LISREL model one must consider the four matrices—LY, BE, PS, and TE. Describing the model parameters residing in them constitutes a major part of constructing the LISREL input file.

Setting up a LISREL Input File

Now the process of constructing the input for a LISREL model can be discussed in detail. As outlined in the flow chart presented in the section Structure of Input Files for SEM Programs, there are three main parts to an input file—data description, model description, and user-specified output.

Every LISREL input file should begin with a title line. For example, using the factor analysis model displayed in Fig. 8, the title line could be

FACTOR ANALYSIS MODEL OF THREE INTERRELATED CONSTRUCTS
EACH MEASURED BY THREE INDICATORS

It is important to note that unlike EQS, the LISREL syntax line does not end with a semicolon.

Next, the data to be analyzed are described in what is referred to as a data-definition line. The data-definition line includes information about the number of variables in the data file, the sample size, and the type of data to be analyzed (e.g., covariance or correlation). (When models are fit to data from more than one group, the number of groups must also be provided in this line; see chap. 6.) Thus, the factor analysis model in Fig. 8 is defined as (using just the first two letters of any keyword):

DA NI=9 NO=245

where DA is the abbreviation for "DAta definition line," NI for "Number of Input variables," and NO for "Number of Observations."

Immediately following this line is the command CM (for Covariance Matrix), followed by the actual covariance matrix to be analyzed.

```
CM
1.01
.32   1.50
.43   .40   1.22
.38   .25   .33   1.13
.30   .20   .30   .7    1.06
.33   .22   .38   .72   .69   1.12
.20   .08   .07   .20   .27   .20   1.30
.33   .19   .22   .09   .22   .12   .69   1.07
.52   .27   .36   .33   .37   .29   .50   .62   1.16
```

Of course, if the data are available as raw data in a separate file, one can refer to the raw data file by just stating RA= followed by the name of the file (e.g., RA=C:\DATA\DATAFILE). Alternatively, if one has already computed the sample covariance matrix and saved it in a separate file, CM= followed by the name of the file can also be used. This completes the first part of the LISREL input file, pertaining to description of the data to be analyzed.

Next come the details concerning the proposed model. This is commonly referred to as the model-definition line. The model-definition line contains information about the number of observed variables (Y) and the number of latent variables (η) in the model. For example, because there are nine observed and three latent variables in Fig. 8, the beginning of the line should read

MO NY=9 NE=3

where MO stands for "MOdel definition line," NY for "Number of Y variables," and NE for "Number of Eta variables." However, this is only the start of the model-description information. As was done in creating the EQS program input file, one must communicate to LISREL information about the model parameters. As it turns out, in order to communicate this information, one must define the status of the four matrices in the general LISREL model (i.e., the matrices LY, BE, PS, and TE; and the information about the nine factor loadings, nine residual variances, and three factor correlations). Of course, in order to describe the confirmatory factor analysis model displayed in Fig. 8, only the matrices LY, PS, and TE are needed because there are no explanatory relationships assumed among the latent variables (i.e., BE is equal to 0).

Based on our experience with the LISREL language, the following status definition of these matrices resembles most cases encountered in practice and permits a full description of models with relatively minimal additional effort. Accordingly, LY is initially defined as a rectangular (full) matrix with fixed elements, and subsequently the model parameters are freed; this definition is accomplished by stating LY=FU,FI, where FU stands for "FUll" and FI for "FIxed" (although this is the default option in LISREL, it is mentioned explicitly to emphasize defining the free factor loadings in the subsequent line). Because there are no explanatory relationships among latent variables, the matrix BE is equal to 0 (i.e., consists only of 0 elements), and because this is also the default option in LISREL, matrix BE is not mentioned in the model statement. Hence, the following model-definition line is used:

MO NY=9 NE=3 LY=FU,FI PS=SY,FR TE=DI,FR

As previously indicated, the factor loading matrix LY is defined as full and fixed (keeping in mind that some of its elements will be freed next). These elements are the factor loadings of the model, which according to Rule 3 are the model parameters. The matrix PS is defined as symmetric and consisting of free parameters (as defined by PS=SY,FR). These parameters are the variances and covariances of the latent variables, which by Rules 1 and 2 are model parameters (some of them will subsequently be fixed, following Rule 6, to set the latent variable metrics). The error covariance matrix TE contains as diagonal elements the remaining model parameters (i.e., the error variances) and is defined as TE=DI,FR (this definition of TE is also a default option in LISREL, but it is included here to emphasize its effect of declaring the error variances to be model parameters.)

With respect to the factor loading parameters, the model definition line has only prepared the grounds for their definition. The complete definition also includes a line that specifically declares the pertinent factor loadings to be free parameters:

FR LY(1, 1) LY(2, 1) LY(3, 1) LY(4, 2) LY(5, 2) LY(6, 2)
FR LY(7, 3) LY(8, 3) LY(9, 3)

or simply

FR LY 1 1 LY 2 1 LY 3 1 LY 4 2 LY 5 2 LY 6 2
FR LY 7 3 LY 8 3 LY 9 3

This definition line starts with FR for "FRee," followed by the notation of the factor loadings in the proposed model declared as free parameters. It is very important to use the previously mentioned double-indexing principle correctly—first state the number of the dependent variable (observed variable in this example), and then the number of the independent variable (latent variable in this example). Because there are nine observed variables and three latent variables, there are potentially, altogether, 27 factor loadings to deal with. However, based on Fig. 8, most of them are 0 because not all variables load on every factor. Instead, every observed variable loads only on its pertinent factor. Declaring the corresponding matrix LY as full and fixed in the model-definition line and then the parameter-freeing line is a quick way to specify the correct factor-loading model parameters.

Up until this point, the factor loadings, independent variable variances and covariances, and error variances have been communicated to LISREL as parameters. That is, all the rules outlined in chapter 1 have been applied, with the exception of Rule 6. According to Rule 6, the metric of each latent variable included in the model must be set. As with EQS, the easiest option is to fix their variances to a value of 1. This metric setting can be achieved by first declaring the factor variances to be fixed parameters (because they were defined in the model-definition line as free), and then assigning the value of 1 to each of them. These two steps are accomplished with the following two input lines:

FI PS(1, 1) PS(2, 2) PS(3, 3)
VA 1 PS(1, 1) PS(2, 2) PS(3, 3)

Thus, one must first FIx the three factor variances, and then assign on the next line the VAlue of 1 to each of them. This finalizes the second part of the LISREL input file and completes the definition of all features of the proposed model.

The final section of the input file is minimal and refers to the kind of output information requested from the LISREL program. There is a simple way of asking for all possible output information from the LISREL program (although with the more complex models, such a request can often lead to an enormous amount of output), and for now this option is used. Thus,

OU ALL

is the final line of a LISREL input file.

The Complete LISREL Input File

The following LISREL input file can be used to examine the proposed model in Fig. 8:

```
FACTOR ANALYSIS MODEL OF THREE INTERRELATED CONSTRUCTS
EACH MEASURED BY THREE INDICATORS
DA NI=9 NO=245
CM
1.01
.32   1.50
.43   .40   1.22
.38   .25   .33   1.13
.30   .20   .30   .7    1.06
.33   .22   .38   .72   .69   1.12
.20   .08   .07   .20   .27   .20   1.30
.33   .19   .22   .09   .22   .12   .69   1.07
.52   .27   .36   .33   .37   .29   .50   .62   1.16
MO NY=9 NE=3 LY=FU,FI PS=SY,FR TE=DI,FR
FR LY(1, 1) LY(2, 1) LY(3, 1) LY(4, 2) LY(5, 2) LY(6, 2)
FR LY(7, 3) LY(8, 3) LY(9, 3)
FI PS(1, 1) PS(2, 2) PS(3, 3)
VA 1 PS(1, 1) PS(2, 2) PS(3, 3)
OU ALL
```

CHAPTER THREE

Path Analysis

WHAT IS PATH ANALYSIS?

Path analysis is an approach to modeling explanatory relationships between observed variables. The explanatory variables are usually assumed to have no measurement error or error that is negligible. The dependent variables may contain errors of measurement, but they are assumed to be present only in the residual terms of the model equations (i.e., that left unexplained by the explanatory variables). A special characteristic of path analysis models is that they do not contain latent variables.

Path analysis has a relatively long history. The term was first used in the early 1900s by the English biometrician Sewell Wright (Wright, 1920, 1921). Wright's approach was developed within a framework that is conceptually similar to the one underlying SEM discussed in chapter 1. The basic idea of path analysis is similar to solving a system of equations resulting when the elements of the sample covariance matrix S are set equal to their counterpart elements of the model reproduced covariance matrix Σ. Wright first demonstrated the application of path analysis to biology by developing models for predicting the birth weight of guinea pigs, examining the relative importance of hereditary influence and environment, and studying human intelligence (Wolfle, 1999). Wright (1921, 1934) provided a useful summary of the approach, complete with methods of calculations and a demonstration of the basic theorem of path analysis by which variable correlations in models can be reproduced by connecting chains of paths. Specifically, Wright's path analysis approach included the following steps. First, write out the model equations relating measured variables. Second, work out the correlations among them in terms of the unknown model parameters. Finally, try to solve the resulting system of equations one at a time, in which the correlations are replaced by the sample correlations, in terms of the parameters.

Using the SEM framework, one can easily fit models conceptualized within the path analysis tradition. This is because path analysis models can be viewed as special cases of structural equation models. Indeed, one can consider any path analysis models as resulting from a corresponding structural model that assumes explanatory relationships between some of its latent variables measured by single indicators with unitary loadings on them, and in which the independent variables involved in those relationships have no residual terms (i.e., no error terms). Hence, to fit a path analysis model one can use a SEM program like EQS or LISREL. Although the application capitalizes on the original idea of fitting models to matrices of interrelationship indices (as developed by Wright, 1934), the actual model-fitting procedure is slightly modified. In particular, even though the independent variables are still considered to be measured without error (as one would do when using the original path analysis approach), the SEM approach considers all model equations simultaneously.

EXAMPLE PATH ANALYSIS MODEL

To demonstrate a path analysis model, consider the following example study (Finn, 1974; see also Jöreskog & Sörbom, 1993b, sec. 4.1.4). The study examined the effects of several variables on university freshmen's academic performance. Five educational measures were collected from a sample of $N = 150$ university freshmen. The following observed variables were used in the study:

1. Grade point average obtained in required courses (AV_REQRD).
2. Grade point average obtained in elective courses (AV_ELECT).
3. High school general knowledge score (SAT).
4. Intelligence score obtained in the last year of high school (IQ).
5. Educational motivation score obtained the last year of high school (ED_MOTIV).

The example path analysis model is presented in Fig. 9. The model is initially presented in EQS notation using V_1 to V_5 for the observed variables, and E_1 and E_2 for the error terms associated with the two dependent variables (i.e., AV_REQRD and AV_ELECT).

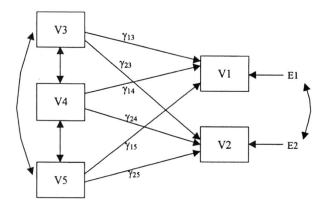

FIG. 9.   Example path analysis model.

In Fig. 9, there are three two-headed "curved arrows" connecting all the independent variables and representing the interrelationships among the high school general knowledge score (SAT), intelligence score (IQ), and motivation score (ED_MOTIV). Note that there are no residuals or measurement errors associated with any of the independent variables. The dependent variables, however, are associated with residual terms. They contain both measurement and prediction error. The curved two-headed arrow connecting the error terms E_1 and E_2 represents their possible interrelation that has been included in the model (it is important to note that including correlated error terms in the path model differs somewhat from the original path analysis method).

The purpose of this path analysis study is to examine the predictive power of several variables on university freshmen's academic performance. In terms of a path analysis, one's interest lies in regressing simultaneously the two dependent variables (V_1 and V_2) on three independent variables (V_3, V_4, and V_5). Note that regressing the two dependent variables is not the same as is traditionally done in a multiple regression analysis, in which a single dependent variable is considered—here the model is a multivariate multiple regression. Thus, in terms of equations, the following relationships are postulated:

$$V_1 = \gamma_{13} V_3 + \gamma_{14} V_4 + \gamma_{15} V_5 + E_1,$$
$$V_2 = \gamma_{23} V_3 + \gamma_{24} V_4 + \gamma_{25} V_5 + E_2, \qquad (11)$$

where γ_{13} to γ_{25} are the six parameters of interest—the partial regression coefficients, also called path coefficients. These path coefficients reflect the predictive strength, in the particular metric used, of SAT, IQ, and ED_MOTIV on the two corresponding dependent variables. Furthermore, in Equations 11 the variables E_1 and E_2 represent the residual terms of the model equations, which reflect not only measurement error but also all additional influences on the pertinent dependent variables, beyond the ones captured by the linear combination of their presumed predictors.

To determine the parameters of the model presented in Fig. 9, one must follow the six rules outlined in chapter 1. Of course, because the model does not deal with latent variables, Rule 6 is not applicable to this model. Using the remaining rules, the parameters are (a) the regression coefficients (i.e., all six γ's in Equations 11), which represent the paths connecting each of the dependent variables (V_1 and V_2) to the predictors (V_3, V_4, and V_5), and (b) the variances and covariances of the independent variables (i.e., the variances and covariances of V_3, V_4, and V_5, as well as the variances and covariance of the equations residual terms E_1 and E_2).

Thus, the model in Fig. 9 has 15 parameters altogether—six path coefficients denoted by γ in Equations 11, six variances and covariances of independent variables, and three variances and covariance of the error terms. Observe that there are no model implications for the variances and covariances of the predictors V_3 through V_5. This is because none of them is a dependent variable. As a result, their variances and covariances are not structured in terms of model parameters. Hence, the covariance matrix of the predictors SAT, IQ, and ED_MOTIV is not restricted, in the sense that the model does not have any consequences with regard to its elements. Therefore, the estimates of these six parameters (three variances and three covariances) will be based entirely on the values in the sample covariance matrix S.

EQS AND LISREL INPUT FILES

EQS Input File

The EQS input file is constructed following the principles outlined in chapter 2. Accordingly, the file begins with a title command line:

```
/TITLE
PATH ANALYSIS MODEL;
```

Next, the number of variables in the model, sample size, and method of estimation are specified:

```
/SPECIFICATIONS
VARIABLES=5; CASES=150;
```

To facilitate interpretation of the outputs, labels are provided for all variables. Using the command line /LABELS, the following names are assigned to each variable in the model (note that variable labels in both EQS and LISREL should not be longer than eight letters).

```
/LABELS
V1=AV_REQRD; V2=AV_ELECT; V3=SAT; V4=IQ; V5=ED_MOTIV;
```

Next the two model-definition equations are provided:

```
/EQUATIONS
V1 = *V3 + *V4 + *V5 + E1;
V2 = *V3 + *V4 + *V5 + E2;
```

followed by the remaining model parameters in the variance and covariance command lines:

```
/VARIANCES
V3 TO V5 = *; E1 TO E2 = *;
/COVARIANCES
V3 TO V5 = *; E1 TO E2 = *;
```

Finally, the data are provided along with the end-of-input file command:

```
/MATRIX
.594
.483    .754
3.993  3.626  47.457
.426   1.757  4.100    10.267
.500   .722   6.394    .525     2.675
/END;
```

The complete EQS input file is as follows (using the appropriate abbreviations):

```
/TIT
PATH ANALYSIS MODEL;
/SPE
VAR=5; CAS=150;
/LAB
V1=AV_REQRD; V2=AV_ELECT; V3=SAT; V4=IQ; V5=ED_MOTIV;
/EQU
V1 = *V3 + *V4 + *V5 + E1;
```

```
V2 = *V3 + *V4 + *V5 + E2;
/VAR
V3 TO V5 = *; E1 TO E2 = *;
/COV
V3 TO V5 = *;
/MAT
.594
.483    .754
3.993   3.626   47.457
.426    1.757   4.100    10.267
.500    .722    6.394    .525     2.675
/END;
```

LISREL Input File

The corresponding LISREL input file is presented next. The program outputs are provided in a separate section.

The LISREL input file requires a slight extension of the general LISREL model notation presented in chapter 2. The discussion is facilitated by the particular symbols used for the path coefficients (partial-regression coefficients) in Equations 11. Accordingly, observed predictor variables are denoted using X, whereas dependent variables remain as Y. That is, V_1 and V_2 now become Y_1 and Y_2 in the LISREL notation, and V_3, V_4, and V_5 become X_1, X_2, and X_3, respectively. The covariance matrix of the predictor variables is referred to as Φ (the Greek letter *phi*), denoted PH in the syntax, and the same principles discussed in chapter 2 apply when referring to its elements. The six regression coefficients in Equations 11 are collected in a matrix Γ (the Greek letter *gamma*), denoted GA in the syntax. The columns of the matrix Γ correspond to the predictors (X_1 to X_3) and its rows are associated with the dependent variables (Y_1 and Y_2). Each entry of the matrix Γ represents a coefficient for the regression of a dependent (Y) on an independent (X) variable. Thus, the elements of the matrix Γ in this example are GA(1,1), GA(1,2), GA(1,3), GA(2,1), GA(2,2), and GA(2,3).

The following LISREL input file is constructed following the principles outlined in chapter 2. The input file also includes some variable labels using the command line LAbels.

```
PATH ANALYSIS MODEL
DA NI=5 NO=150
CM
.594
.483    .754
3.993   3.626   47.457
```

```
   .426   1.757  4.100  10.267
   .500    .722  6.394   .525    2.675
   LA
   AV_REQRD AV_ELECT SAT IQ ED_MOTIV
   MO NY=2 NX=3 GA=FU,FR PH=SY,FR PS=SY,FR
   OU
```

Using the same title, the data-definition line declares that the model will be fit to data on five variables collected from 150 subjects. The sample covariance matrix signaled by CM is provided next, along with the variable labels. In the model command line, the notation NX denotes the "Number of X variables." The matrix GA is, accordingly, declared to be full of free model parameters—these are the six γ's in Equations 11. The (symmetric) covariance matrix of the predictors (i.e., PH) is defined as containing model free parameters (as was done when creating the EQS input file). The elements of the PH matrix correspond to the variances and covariances of SAT, IQ, and ED_MOTIV variables. Finally, the covariance matrix of the residual terms of the dependent variables Y_1 and Y_2 (i.e., the matrix PS) is also defined as being symmetric and containing free model parameters. The elements of the PS matrix correspond to the variances and covariance of the two equations' error terms.

MODELING RESULTS

EQS Program Results

The output produced by the EQS input file created in the previous section is presented next. We introduce at this point the convention of displaying the output results in a different and proportionate font, so that they stand out from the main text in the book. In addition, at appropriate places comments are inserted clarifying portions of the output (usually approximately at the end of each page of the output) and occasionally annotate the output.

The EQS output begins with

```
PROGRAM CONTROL INFORMATION

  1   /TITLE
  2   PATH ANALYSIS MODEL;
  3   /SPECIFICATIONS
  4   VARIABLES=5; CASES=150; METHOD=ML;
  5   /LABELS
  6   V1=AV_REQRD; V2=AV_ELECT; V3=SAT; V4=IQ; V5=ED_MOTIV;
  7   /EQUATIONS
```

```
 8   V1 = *V3 + *V4 + *V5 + E1;
 9   V2 = *V3 + *V4 + *V5 + E2;
10   /VARIANCES
11   V3 TO V5 = *; E1 TO E2 = *;
12   /COVARIANCES
13   V3 TO V5 = *; E1 TO E2 = *;
14   /MATRIX
15   .594
16   .483    .754
17   3.993   3.626    47.457
18   .426    1.757    4.100    10.267
19   .500    .722     6.394    .525     2.675
20   /END;
```

```
20 RECORDS OF INPUT MODEL FILE WERE READ
```

This part of the output merely echoes the input file submitted to EQS and any mistakes made while creating the input can be easily spotted. It is quite important, therefore, that this first section of the output always be carefully examined before looking at the other output sections.

```
COVARIANCE MATRIX TO BE ANALYZED: 5 VARIABLES (SELECTED FROM 5 VARIABLES)
BASED ON 150 CASES.
```

		AV_REQRD	AV_ELECT	SAT	IQ	ED_MOTIV
		V 1	V 2	V 3	V 4	V 5
AV_REQRD	V 1	0.594				
AV_ELECT	V 2	0.483	0.754			
SAT	V 3	3.993	3.626	47.457		
IQ	V 4	0.426	1.757	4.100	10.267	
ED_MOTIV	V 5	0.500	0.722	6.394	0.525	2.675

```
BENTLER-WEEKS STRUCTURAL REPRESENTATION:

        NUMBER OF DEPENDENT VARIABLES =  2
              DEPENDENT V'S:      1    2

        NUMBER OF INDEPENDENT VARIABLES =  5
              INDEPENDENT V'S:      3    4    5
              INDEPENDENT E'S:      1    2

  3RD STAGE OF COMPUTATION REQUIRED     1410 WORDS OF MEMORY.
  PROGRAM ALLOCATE      50000 WORDS

  DETERMINANT OF INPUT MATRIX IS      0.26607E+02
```

In the second section of the output EQS provides information about the details of the model (i.e., about the number of independent and dependent variables, and the covariance matrix actually analyzed), as well as details relating to the internal organization of the memory to accomplish the computational routine. The bottom of this second output page also contains a message that can be particularly important for detecting numerical difficulties. The message has to do with the DETERMINANT OF THE INPUT MATRIX, which is, in very simple terms, a number that reflects the generalized variance of the 5 variables. If the determinant (variance) is 0, matrix computations cannot be conducted (i.e., in matrix-algebra terminology, the covariance matrix is singular). In cases where the determinant is very close to 0, matrix computations can be unreliable and any obtained numerical solutions are quite unstable. In common statistical terminology, the presence of a determinant close to 0 is a clue that there is a problem of a nearly perfect linear dependency (i.e., multicollinearity) among the observed variables analyzed. For example, in a regression analysis the presence of multicollinearity implies that one is using redundant information in the regression model, which can easily lead to an artificial inflation of the prediction model because, in a sense, this redundant information is used more than once. Of course, a simple solution may be to just drop an offending variable that is linearly related to the other variables and respecify the model. Thus, examining the determinant of the matrix in the EQS output section provides important clues about the accuracy of the data analysis.

```
MAXIMUM LIKELIHOOD SOLUTION (NORMAL DISTRIBUTION THEORY)

PARAMETER ESTIMATES APPEAR IN ORDER,
NO SPECIAL PROBLEMS WERE ENCOUNTERED DURING OPTIMIZATION.
```

This is also a very important message. It indicates that the program has not encountered problems stemming from lack of model identification. Otherwise, this is the place where one would see a warning message entitled CONDITION CODE that would indicate which parameters are possibly unidentified. For this model, the NO SPECIAL PROBLEMS message is a reassurance that the model is technically sound and identified.

```
RESIDUAL COVARIANCE MATRIX (S-SIGMA):

                  AV_REQRD   AV_ELECT      SAT        IQ      ED_MOTIV
                   V   1      V   2       V   3      V   4       V   5
AV_REQRD V   1     0.000
AV_ELECT V   2     0.000      0.000
   SAT   V   3     0.000      0.000       0.000
   IQ    V   4     0.000      0.000       0.000     0.000
ED_MOTIV V   5     0.000      0.000       0.000     0.000      0.000

                 AVERAGE ABSOLUTE COVARIANCE RESIDUALS      =      0.0000
       AVERAGE OFF-DIAGONAL ABSOLUTE COVARIANCE RESIDUALS   =      0.0000
```

STANDARDIZED RESIDUAL MATRIX:

		AV_REQRD	AV_ELECT	SAT	IQ	ED_MOTIV
		V 1	V 2	V 3	V 4	V 5
AV_REQRD V	1	0.000				
AV_ELECT V	2	0.000	0.000			
SAT V	3	0.000	0.000	0.000		
IQ V	4	0.000	0.000	0.000	0.000	
ED_MOTIV V	5	0.000	0.000	0.000	0.000	0.000

```
            AVERAGE ABSOLUTE STANDARDIZED RESIDUALS      =      0.0000
    AVERAGE OFF-DIAGONAL ABSOLUTE STANDARDIZED RESIDUALS  =      0.0000
```

The RESIDUAL COVARIANCE MATRIX contains the computed variable variance and covariance residuals. As discussed in chapter 1, these are the differences between the counterpart elements of the empirical covariance matrix S (given in the last input section /MATRIX) and the one reproduced by the model at the final solution (i.e., Σ). Thus, the values in the residual covariance matrix correspond to the differences between the two matrices (i.e., $S - \Sigma$). In this sense, the residual matrix is a complex measure of model fit for the variances and covariances, as opposed to a single number provided by most fit indices. Large standardized residuals are indicative of possibly serious deficiencies of the model in that particular part (i.e., with regard to the specific variable variance or covariance).

In the present example, there are no non-0 residuals. On the contrary, in this study there is a perfect fit of the model to the data—the model perfectly reproduces the analyzed covariance matrix because all residuals are 0. As it turns out, the reason for this finding is that we are dealing with a saturated or just-identified model (i.e., one that has as many parameters as there are nonredundant elements of the covariance matrix). Recall that the model has 15 parameters and, with five observed variables, there are $p(p + 1)/2 = 5(5 + 1)/2 = 15$ nonredundant elements in the analyzed covariance matrix. Thus, the model examined in this study has as many parameters as there are nonredundant elements in the covariance matrix. Because saturated models like this one will always fit the data perfectly, there is no way one can really test or confirm the plausibility of the model (see chap. 1 for a complete discussion).

LARGEST STANDARDIZED RESIDUALS:

```
    V   2,V  1      V   4,V  2      V   5,V  2      V   3,V  2      V   1,V  1
      0.000           0.000           0.000           0.000           0.000
    V   3,V  3      V   4,V  1      V   2,V  2      V   4,V  3      V   4,V  4
      0.000           0.000           0.000           0.000           0.000
    V   5,V  1      V   3,V  1      V   5,V  3      V   5,V  4      V   5,V  5
      0.000           0.000           0.000           0.000           0.000
```

*DISTRIBUTION OF STANDARDIZED RESIDUALS

```
    - - - - - - - - - - - - - - - - - - - - - - - - -
    !                                    !
 20-                                     -
    !                                    !
    !                                    !
    !                                    !
    !                                    !         RANGE        FREQ  PERCENT
 15-                                     -
    !                   *                !  1  -0.5  -  --         0    0.00%
    !                   *                !  2  -0.4  - -0.5        0    0.00%
    !                   *                !  3  -0.3  - -0.4        0    0.00%
    !                   *                !  4  -0.2  - -0.3        0    0.00%
 10-                    *                -  5  -0.1  - -0.2        0    0.00%
    !                   *                !  6   0.0  - -0.1       14   93.33%
    !                   *                !  7   0.1  -  0.0        1    6.67%
    !                   *                !  8   0.2  -  0.1        0    0.00%
    !                   *                !  9   0.3  -  0.2        0    0.00%
  5-                    *                -  A   0.4  -  0.3        0    0.00%
    !                   *                !  B   0.5  -  0.4        0    0.00%
    !                   *                !  C   ++   -  0.5        0    0.00%
    !                   *                !   - - - - - - - - - - - - - - - - - -
    !                   *  *             !       TOTAL            15  100.00%
    - - - - - - - - - - - - - - - - - - - - - - -
         1  2  3  4  5  6  7  8  9  A  B  C   EACH "*" REPRESENTS 1 RESIDUALS
```

This section of the output provides only rearranged information about the fit of the model. Of course, in the case of a less-than-perfect fit, a fair amount of information about the model can be obtained from this section. For example, the upper part of the output section provides a convenient summary of where to find the largest residuals (i.e., the largest deviations of the model from the data). The lower part of the output section provides information about the distribution of the residuals. In this example, due to the numerical estimation involved in fitting the model, the obtained residuals are not perfectly equal to 0 out to all the decimal places used by the program. This is the reason there is a spike in the center of the distribution and that one residual happens to fall off. With well-fitting models one should expect all residuals to be small and concentrated in the central part of the distribution of asterisks symbolizing the standardized residuals, and the distribution to be in general symmetric.

MAXIMUM LIKELIHOOD SOLUTION (NORMAL DISTRIBUTION THEORY)

GOODNESS OF FIT SUMMARY

```
INDEPENDENCE MODEL CHI-SQUARE =       460.156 ON    10 DEGREES OF FREEDOM
INDEPENDENCE AIC =    440.15574    INDEPENDENCE CAIC =    400.04938
        MODEL AIC =      0.00002          MODEL CAIC =      0.00002
```

```
CHI-SQUARE =        0.000 BASED ON      0 DEGREES OF FREEDOM
NONPOSITIVE DEGREES OF FREEDOM. PROBABILITY COMPUTATIONS ARE UNDEFINED.

BENTLER-BONETT NORMED    FIT INDEX=       1.000

NON-NORMED FIT INDEX WILL NOT BE COMPUTED BECAUSE A DEGREES OF FREEDOM IS
ZERO.
```

In this next output section, several goodness-of-fit indices are presented. The first feature to note concerns the 0 degrees of freedom for the model. As discussed previously, this is a consequence of the fact that the model has as many parameters as there are nonredundant elements in the observed covariance matrix. The chi-square value is also 0, indicating, again, perfect fit. This result is not surprising because the model fits the covariance matrix perfectly. In contrast, the chi-square value of the independence model is quite large. This result is also expected because this model assumes no relationships between the variables, and hence represents in general a poor means of describing the analyzed data. In fact, large chi-square values for the independence model are quite frequent in practice—particularly when the variables of interest exhibit some interrelationship. The Akaike information criterion (AIC) and its consistent-estimator version (CAIC) are practically 0 for the fitted model, and much smaller than those of the independence model. This finding is also explained by the fact that the independence model is a poor means of describing the analyzed data, whereas the one we fitted is far better. Finally, because a saturated model was fit in this example, which by necessity has both a 0 chi-square value and 0 degrees of freedom, the Bentler–Bonett nonnormed fit index cannot be computed. The reason is that, by definition, the Bentler–Bonett index involves division by the degrees of freedom of the proposed model (which here are 0). In contrast, the Bentler–Bonett normed fit index is 1, which is its maximum possible value typically associated with a perfect fit.

```
                        ITERATIVE SUMMARY


                      PARAMETER
ITERATION             ABS CHANGE       ALPHA          FUNCTION
    1                 14.688357       1.00000         5.14968
    2                 14.360452       1.00000         0.00000
    3                  0.000000       1.00000         0.00000
```

The iterative summary provides an account of the numerical routine performed by EQS to minimize the maximum likelihood fit function (matrix distance) used in accordance with the multinormality of the data. It took the program three iterations to find the final solution. This was achieved after the fit function quickly diminished to 0. Because this is a saturated model with perfect fit, the absolute minimum of 0 is achieved by the fit function, which is not generally the case with nonsaturated models, as is demonstrated in later examples.

```
MEASUREMENT EQUATIONS WITH STANDARD ERRORS AND TEST STATISTICS

AV_REQRD=V1   =      .086*V3      +   .008*V4    + -.021*V5    + 1.000 E1
                     .007             .013            .031
                  11.642             .616           -.676


AV_ELECT=V2   =      .046*V3      +   .146*V4    +  .131*V5    + 1.000 E2
                     .007             .013            .030
                   6.501           11.560          4.434
```

This is the final solution, presented in nearly the same form as the model equations provided to EQS in the input file. However, in this output the asterisks are preceded by the computed parameter estimates. (It is important to emphasize that the asterisks are used throughout EQS input and output to denote estimated parameters, and do not have any relation to significance.) Immediately beneath each parameter estimate is its standard error. The standard errors are measures of estimate stability (i.e., precision of estimation). Dividing each parameter estimate by its standard error yields what is referred to as a *t* value and this information is provided beneath the standard error. As mentioned in chapter 1, if the *t* value is outside the (–2; +2) range one can suggest that the pertinent parameter is likely to be non-0 in the population; otherwise, it can be treated as 0 in the population. The *t* value, therefore, represents a simple test statistic of the null hypotheses that the corresponding model parameter is 0 in the population.

As can be seen in the output section, the *t* values indicate that all path coefficients are significant, except the impacts of IQ and ED_MOTIV on AV_REQRD, which are associated with nonsignificant *t* values (inside the –2; +2 interval). Thus, IQ and ED_MOTIV seem to be unimportant variables in the population. That is, there appears to be only weak evidence that IQ and ED_MOTIV might matter for student performance in required courses (AV_REQRD) once the impact of SAT is accounted for (this example is reconsidered later in the chapter with more precise statements).

```
VARIANCES OF INDEPENDENT VARIABLES
----------------------------------
                          V                         F
                          ---                       ---
V3  -  SAT                47.457*I                    I
                           5.498 I                    I
                           8.631 I                    I
                                 I                    I
V4  -  IQ                 10.267*I                    I
                           1.190 I                    I
                           8.631 I                    I
                                 I                    I
```

```
V5 -ED_MOTIV                    2.675*I                        I
                                 .310 I                        I
                                8.631 I                        I
                                      I                        I
```

These are the estimated variances of independent manifest variables along with their standard errors and *t* values (i.e., for SAT, IQ, and ED_MOTIV).

```
VARIANCES OF INDEPENDENT VARIABLES
-------------------------------

                         E                            D
                        ---                          ---

E1 -AV_REQRD                    .257*I                        I
                                .030 I                        I
                               8.631 I                        I
                                     I                        I
E2 -AV_ELECT                    .236*I                        I
                                .027 I                        I
                               8.631 I                        I
                                     I                        I
```

These are the variances of the residual terms along with their standard errors and *t* values. As are the variances of the predictor variables, they are all significant.

```
COVARIANCES AMONG INDEPENDENT VARIABLES
-------------------------------------.

                         V                            F
                        ---                          ---

V4 -  IQ                       4.100*I                        I
V3 -  SAT                      1.839 I                        I
                               2.229 I                        I
                                     I                        I
V5 -ED_MOTIV                   6.394*I                        I
V3 -  SAT                      1.061 I                        I
                               6.025 I                        I
                                     I                        I
V5 -ED_MOTIV                    .525*I                        I
V4 -  IQ                        .431 I                        I
                               1.217 I                        I
                                     I                        I


COVARIANCES AMONG INDEPENDENT VARIABLES
-------------------------------------.

                         E                            D
                        ---                          ---

E2 -AV_ELECT                    .171*I                        I
E1 -AV_REQRD                    .025 I                        I
                               6.971 I                        I
                                     I                        I
```

These are the covariances among the predictor variables along with their standard errors and t values.

```
STANDARDIZED SOLUTION:

AV_REQRD=V1  =   .771*V3    + .034*V4   + -.044*V5   + .657 E1
AV_ELECT=V2  =   .366*V3    + .539*V4   + .247*V5    + .559 E2
```

The STANDARDIZED SOLUTION output results from standardizing all variables in the model. The standardized solution uses a metric that is uniform across all measures and, hence, makes possible some assessment of the relative importance of the predictors. Another way to use this information involves squaring the coefficients associated with the error terms. The resulting values provide information about the percentage of unexplained variance in the dependent variables. In a sense, these squared values are analogs to $1 - R^2$ indices corresponding to regression models for each equation. As can be seen in this example, some 43% (= 0.657^2 in percentage) of individual differences in AV_REQRD could not be predicted by SAT, IQ, and ED_MOTIV. Similarly, some 31% of individual differences in AV_ELECT were not explained by SAT, IQ, and ED_MOTIV.

```
CORRELATIONS AMONG INDEPENDENT VARIABLES
------------------------------------.

                      V                        F
                      ---                      ---
V4 - IQ             .186*I                      I
V3 - SAT              I                         I
                      I                         I
V5 -ED_MOTIV        .567*I                      I
V3 - SAT              I                         I
                      I                         I
V5 -ED_MOTIV        .100*I                      I
V4 - IQ               I                         I
                      I                         I


CORRELATIONS AMONG INDEPENDENT VARIABLES
------------------------------------.

                      E                        D
                      ---                      ---
E2 -AV_ELECT        .696*I                      I
E1 -AV_REQRD          I                         I
                      I                         I
--------------------------------------------------------------------
                E N D   O F   M E T H O D
--------------------------------------------------------------------
```

Finally, at the end of the EQS output, the correlations of all the independent model variables are presented. These values are simply reexpressions of earlier parts of the output dealing with estimated independent variances and covariances. The magnitude of the correlation between the two equations' error terms suggests that the unexplained portions of individual differences in AV_ELECT and AV_REQRD are quite high. One may suspect that this may be a consequence of common omitted variables. This possibility can be addressed by a subsequent and more comprehensive study that includes as predictors variables in addition to the SAT, IQ, and ED_MOTIV used in this example.

LISREL Program Results

The LISREL input described previously produces the following results rounded off to two decimals, which is the default option in LISREL (unlike EQS, in which the default is three). Once again, in presenting the output sections of the LISREL program, the font is changed, comments are inserted at appropriate places, and the logo of the program and its recurring first title line are dispensed with in order to save space.

```
The following lines were read from file PA.LSR:

PATH ANALYSIS MODEL
DA NI=5 NO=150
CM
.594
.483 .754
3.993 3.626 47.457
.426 1.757 4.1 10.267
.500 .722 6.394 .525 2.675
LA
AV_REQRD AV_ELECT SAT IQ ED_MOTIV
MO NY=2 NX=3 GA=FU,FR PH=SY,FR PS=SY,FR
OU
```

As in EQS, the LISREL output first echoes the input file. This is very useful for checking whether the model actually fitted is indeed the one intended to be analyzed by the researcher.

```
                NUMBER OF INPUT VARIABLES    5
                NUMBER OF Y - VARIABLES      2
                NUMBER OF X - VARIABLES      3
                NUMBER OF ETA - VARIABLES    2
                NUMBER OF KSI - VARIABLES    3
                NUMBER OF OBSERVATIONS     150
```

Next, a summary of the variables in the model is given, in terms of observed, unobserved variables, and sample size. This section is also quite useful for checking whether the number of variables in the model have been correctly specified. Although in this example, the number of ETA and

KSI variables are irrelevant, they will prove to be quite important later when dealing with the full LISREL model (see chap. 5).

COVARIANCE MATRIX TO BE ANALYZED

	AV_REQRD	AV_ELECT	SAT	IQ	ED_MOTIV
	--------	--------	--------	--------	--------
AV_REQRD	.59				
AV_ELECT	.48	.75			
SAT	3.99	3.63	47.46		
IQ	.43	1.76	4.10	10.27	
ED_MOTIV	.50	.72	6.39	.53	2.68

The covariance matrix provided in the LISREL input file is also echoed in the output and should be examined for potential errors.

PARAMETER SPECIFICATIONS

GAMMA

	SAT	IQ	ED_MOTIV
	--------	--------	--------
AV_REQRD	1	2	3
AV_ELECT	4	5	6

PHI

	SAT	IQ	ED_MOTIV
	--------	--------	--------
SAT	7		
IQ	8	9	
ED_MOTIV	10	11	12

PSI

	AV_REQRD	AV_ELECT
	--------	--------
AV_REQRD	13	
AV_ELECT	14	15

This is a very important section that identifies the model parameters as declared in the input file, and then numbers them consecutively. Each free model parameter is assigned a separate number. It is important to note that all parameters that are constrained to be equal to one another are given the same number, whereas parameters that are fixed are not numbered (i.e., they are assigned 0; see discussion in next section). According to this section output, LISREL understood that the model has 15 parameters altogether (see Fig. 9). These include the six regression (path) coefficients in the GAMMA matrix that relate each of the predictors to the dependent variables (with predictor variables listed as columns and dependent variables as rows of the matrix); the PHI matrix, containing all six variances and covariances of the predictors as parameters; and the PSI matrix, containing the variances and covariances of the residual terms of the dependent variables (because both the PHI and PSI matrices are symmetric, only the elements along the diagonal and beneath it are nonredundant, and hence only they are assigned consecutive numbers).

```
LISREL ESTIMATES (MAXIMUM LIKELIHOOD)

        GAMMA
                    SAT           IQ      ED_MOTIV
                 --------     --------     --------
AV_REQRD            .09          .01         -.02
                  (.01)        (.01)        (.03)
                 11.52          .61         -.67

AV_ELECT            .05          .15          .13
                  (.01)        (.01)        (.03)
                  6.44        11.44         4.39

        COVARIANCE MATRIX OF Y AND X

                 AV_REQRD     AV_ELECT          SAT           IQ      ED_MOTIV
                 --------     --------     --------     --------     --------
AV_REQRD            .59
AV_ELECT            .48          .75
     SAT           3.99         3.63        47.46
      IQ            .43         1.76         4.10        10.27
ED_MOTIV            .50          .72         6.39          .53         2.68

        PHI
                    SAT           IQ      ED_MOTIV
                 --------     --------     --------
     SAT          47.46
      IQ           4.10        10.27
ED_MOTIV           6.39          .53         2.68

        PSI
                 AV_REQRD     AV_ELECT
                 --------     --------
AV_REQRD            .26
                  (.03)
                  8.54

AV_ELECT            .17          .24
                  (.02)        (.03)
                  6.90         8.54

        SQUARED MULTIPLE CORRELATIONS FOR STRUCTURAL EQUATIONS
                 AV_REQRD     AV_ELECT
                 --------     --------
                    .57          .69
```

This is the final output solution provided by LISREL. In a way similar to the EQS program output, LISREL presents the parameter estimates along with their standard errors and *t* values in a column format. Note that the matrix presented in this section under the title COVARIANCE MATRIX OF

Y AND X is actually the final model reproduced covariance matrix Σ. As can be seen, in this case Σ is identical to the sample covariance matrix S because the model is saturated and hence fits or reproduces it perfectly. Note also that the entries in the covariance matrix PHI of the independent variables are not associated with standard errors, unlike the other parameter estimates. This is to emphasize that they are identical to the corresponding elements of the sample covariance matrix (see lower-right corner of the empirical covariance matrix in the input file). The reason is that the model does not have any implications for the variances and covariances for the three predictor variables SAT, IQ, and ED_MOTIV. As a result, they are directly estimated at the corresponding empirical values of the sample covariance matrix S.

The section ends by providing the squared multiple correlations for the two structural equations of the model, one per dependent variable. The squared multiple correlations provide information about the percentage of explained variance in the dependent variables. Thus, in a sense they are analogs to the R^2 indices corresponding to regression models for each equation (and are the same up to rounding-off error as those obtained using the EQS output).

```
                  GOODNESS OF FIT STATISTICS
   CHI-SQUARE WITH 0 DEGREE OF FREEDOM = 0.00 (P = 1.00)
          The Model is Saturated, the Fit is Perfect !
```

The goodness-of-fit statistics output is yet another indication of the perfect fit of the model (i.e., due to the fact that the model is saturated). It is important to note that with most models examined in practice (which are nonsaturated models), a researcher should first inspect the goodness-of-fit statistics provided in this section of the output before interpreting any parameter estimates. Using this strategy ensures that a researcher's interpretation of parameter estimates is carried out only for models that are reasonable approximations of the analyzed data. For models that are rejected as means of data representation, interpretations of parameter estimates should not be generally carried out because they can yield substantively misleading results.

```
MODIFICATION INDICES AND EXPECTED CHANGE
NO NON-ZERO MODIFICATION INDICES FOR GAMMA
NO NON-ZERO MODIFICATION INDICES FOR PSI
```

Finally, the reported lack of non-0 modification indices is another indication that there is no way to improve the model (i.e., because the model already fits the analyzed data perfectly). As it turns out, however, there will be situations in practice when model modifications must be made in order to achieve a better model fit or to address further substantively meaningful questions. The technical aspects of considering such models are discussed in detail in the next section.

TESTING MODEL RESTRICTIONS IN SEM

The model examined in the previous section was a very special one—it was a saturated or just-identified model. As a result, the model fit the data perfectly (something that one cannot expect to see with most models considered in practice). Because the model was saturated, it had 0 degrees of freedom and could not really be tested. This is because models with 0 degrees of freedom can never be disproved when tested against observed data, regardless of their theoretical implications or how that data looks. As a consequence, saturated models are generally not very interesting in substantive research. However, a saturated model can provide a very useful baseline against which other models with positive degrees of freedom can be tested. This is the reason why a saturated model was examined first.

Issues of Fit with Nested Models

The saturated model presented in the previous section and displayed in Fig. 9 also facilitates the introduction of a very important topic in SEM testing—that of a nested model. A model M is said to be *nested* in another model M′ if M can be obtained from M′ by fixing one or more of its parameters to 0 or some other constant(s), or to fulfil a particular relationship (e.g., a linear combination of them to equal 0 or another constant). Actually, there also seems to be an informal tendency in the literature to refer to such pairs of models as nested models. In fact, there is potentially no limit to the number of nested models in a considered sequence, or to the number of parameters that are fixed or constrained in one of them in order to obtain the more restricted model(s).

Another interesting aspect about a nested model is that it will always have a greater number of degrees of freedom than the model in which it is nested (recall that the nested model M is obtained from M′ by constraining one or more parameters to 0 or other constant(s), or to fulfil a particular relationship). As it turns out, the difference between the chi-squares of the two models, based on their degrees of freedom, can be used as a test if this restriction imposed on one or more parameters resulted in a significant decrement in fit. The test is generally referred to as a chi-square difference test and is based on a chi-square distribution with degrees of freedom equal to the difference between the degrees of freedom of the models compared.

In practice, researchers typically consider only those nested models that impose substantively interesting restrictions on parameters. Thus, by comparing the fit of models and taking into account the difference in the number of their parameters, researchers examine the plausibility of the imposed restrictions. In general, the more restricted model (being a constrained version of another model) will typically be associated with a higher chi-square value because it imposes more restrictions on the model parameters. These restrictions, in effect, make the model less able to emulate the analyzed covariance matrix, and hence the resulting difference in the chi-square values will be positive. Similarly, the difference in degrees of freedom obtained when the degrees of freedom of the more general model are subtracted from those of the more restricted one is also positive because the more restricted model has fewer parameters and hence more degrees of freedom.

It can be shown that with large samples, the difference in chi-square values of nested models follows approximately a chi-square distribution with degrees of freedom equal to the difference in their degrees of freedom. For simplicity, the difference in chi-square values is denoted by ΔT, and the difference between the degrees of freedom by Δdf. If the ΔT is significant when assessed against the chi-square distribution with Δdf degrees of freedom, it is an indication that the imposed restrictions in the more constrained model resulted in a significant decrement in model fit. Conversely, if the ΔT is nonsignificant, one can retain the hypothesis concerning the plausibility of the imposed constraints. It is important to note that model comparison can also proceed in the other direction—models can have parameters sequentially freed, and the successive models can be compared to assess whether freeing a particular parameter resulted in a significant increment in model fit. Thus, in general terms the chi-square difference test can be considered analogous to the test of change in R^2 when adding predictors to (or removing them from) a set of explanatory variables in regression analysis, or equivalently to the F test for considering dropping predictors in a regression equation.

Testing Restrictive Hypotheses in the Example

Returning to the specific model presented in Fig. 9, the hypothesis can be examined that intelligence (IQ) and motivation (ED_MOTIV) have the same impact on grade point average obtained in elective subjects (AV_ELECT). To accomplish this, the equality restriction on the paths leaving IQ and ED_MOTIV and ending at AV_ELECT is introduced. This restriction can be included in the EQS input file by adding a /CONSTRAINT section, and in the LISREL file by adding an input line that imposes the particular parameters' identity. The necessary changes in the EQS file are as follows (note the extended title, indicating the additional restriction feature of this model):

```
/TITLE
PATH ANALYSIS MODEL WITH IMPOSED CONSTRAINTS;
/SPECIFICATIONS
VARIABLES=5; CASES=150;
/LABELS
V1=AV_REQRD; V2=AV_ELECT; V3=SAT; V4=IQ; V5=ED_MOTIV;
/EQUATIONS
V1 = *V3 + *V4 + *V5 + E1;
V2 = *V3 + *V4 + *V5 + E2;
/VARIANCES
V3 TO V5 = *; E1 TO E2 = *;
/COVARIANCES
V3 TO V5 = *;
/CONSTRAINT
(V2,V4) = (V2,V5); ! THIS IS THE NECESSARY RESTRICTION;
/MATRIX
.594
.483    .754
3.993   3.626   47.457
.426    1.757   4.100    10.267
.500    .722    6.394    .525     2.675
/END;
```

As indicated by the added comment in the constraint input line, the restriction under consideration is imposed by requiring the identity of the path coefficients of the involved variables.

In the corresponding LISREL input file the parameter restriction is indicated by using an EQuality statement that sets the two involved parameters equal to one another as follows:

```
PATH ANALYSIS MODEL WITH IMPOSED CONSTRAINTS
DA NI=5 NO=150
CM
.594
.483    .754
3.993   3.626   47.457
.426    1.757   4.100    10.267
.500    .722    6.394    .525     2.675
LA
AV_REQRD AV_ELECT SAT IQ ED_MOTIV
```

```
MO NY=2 NX=3 GA=FU,FR PH=SY,FR PS=SY,FR
EQ GA(2, 2) GA(2, 3) ! THIS IS THE NECESSARY RESTRICTION
OU
```

Thus, the EQuality line sets the two considered path coefficients equal in the restricted model.

Because the results of the two programs are identical, only the EQS output file is discussed here (interested readers may verify this by running the corresponding LISREL input file). In addition, the first two pages of the EQS output, which are identical to those of the model without the constraint are not presented.

```
MAXIMUM LIKELIHOOD SOLUTION (NORMAL DISTRIBUTION THEORY)

PARAMETER ESTIMATES APPEAR IN ORDER,
NO SPECIAL PROBLEMS WERE ENCOUNTERED DURING OPTIMIZATION.
ALL EQUALITY CONSTRAINTS WERE CORRECTLY IMPOSED
```

As indicated before, this message provided by the EQS program is important because it indicates that the program has not encountered any problems stemming from lack of model identification and that the model constraints were correctly introduced.

```
RESIDUAL COVARIANCE MATRIX (S-SIGMA):
```

		AV_REQRD	AV_ELECT	SAT	IQ	ED_MOTIV
		V 1	V 2	V 3	V 4	V 5
AV_REQRD	V 1	0.000				
AV_ELECT	V 2	0.000	0.000			
SAT	V 3	0.000	0.000	0.000		
IQ	V 4	0.017	0.023	0.000	0.000	
ED_MOTIV	V 5	-0.017	-0.023	0.000	0.000	0.000

```
                AVERAGE ABSOLUTE COVARIANCE RESIDUALS       =       0.0054
        AVERAGE OFF-DIAGONAL ABSOLUTE COVARIANCE RESIDUALS   =       0.0080


STANDARDIZED RESIDUAL MATRIX:
```

		AV_REQRD	AV_ELECT	SAT	IQ	ED_MOTIV
		V 1	V 2	V 3	V 4	V 5
AV_REQRD	V 1	0.001				
AV_ELECT	V 2	0.000	0.000			
SAT	V 3	0.000	0.000	0.000		
Q	V 4	0.007	0.008	0.000	0.000	
ED_MOTIV	V 5	-0.013	-0.016	0.000	0.000	0.000

```
                AVERAGE ABSOLUTE STANDARDIZED RESIDUALS       =       0.0031
        AVERAGE OFF-DIAGONAL ABSOLUTE STANDARDIZEDCO RESIDUALS =       0.0045
```

LARGEST STANDARDIZED RESIDUALS:

V 5,V 2	V 5,V 1	V 4,V 2	V 4,V 1	V 1,V 1
-0.016	-0.013	0.008	0.007	0.001

V 2,V 1	V 5,V 5	V 5,V 4	V 2,V 2	V 5,V 3
0.000	0.000	0.000	0.000	0.000

V 4,V 4	V 4,V 3	V 3,V 2	V 3,V 1	V 3,V 3
0.000	0.000	0.000	0.000	0.000

DISTRIBUTION OF STANDARDIZED RESIDUALS

```
      - - - - - - - - - - - - - - - - - - - - - - - - - - - -
      !                                    !
 20-  !                                    -
      !                                    !
      !                                    !
      !                                    !
      !                                    !    RANGE      FREQ  PERCENT
 15-  !                                    -
      !                                    !  1  -0.5  -  --      0    0.00%
      !                                    !  2  -0.4  - -0.5     0    0.00%
      !                                    !  3  -0.3  - -0.4     0    0.00%
      !                                    !  4  -0.2  - -0.3     0    0.00%
 10-  !                                    -  5  -0.1  - -0.2     0    0.00%
      !                                    !  6   0.0  - -0.1     7   46.67%
      !               *                    !  7   0.1  -  0.0     8   53.33%
      !            *  *                     !  8   0.2  -  0.1     0    0.00%
      !            *  *                     !  9   0.3  -  0.2     0    0.00%
  5-  !            *  *                     -  A   0.4  -  0.3     0    0.00%
      !            *  *                     !  B   0.5  -  0.4     0    0.00%
      !            *  *                     !  C   ++   -  0.5     0    0.00%
      !            *  *                     !  - - - - - - - - - - - - - - - - - -
      !            *  *                     !     TOTAL          15  100.00%
      - - - - - - - - - - - - - - - - - - - - - - - - - - - -
          1  2  3  4  5  6  7  8  9  A  B  C   EACH "*" REPRESENTS  1 RESIDUALS
```

As expected, the constrained model does not perfectly reproduce the observed covariance matrix (as was the case with the saturated model). This is because most models fit in research are not perfect representations of the analyzed data and therefore do not completely reproduce the analyzed matrices of variable interrelationships. In the particular case under consideration, the largest standardized residuals are, however, fairly small (note also their symmetric distribution around 0 in the latter part of the output portion). This result suggests that the model fits all elements of the sample covariance matrix well.

GOODNESS OF FIT SUMMARY

INDEPENDENCE MODEL CHI-SQUARE =   460.156 ON 10 DEGREES OF FREEDOM

```
INDEPENDENCE AIC =    440.15574   INDEPENDENCE CAIC    =   400.04938
          MODEL AIC =    -1.77898          MODEL CAIC    =    -5.78961

CHI-SQUARE =        0.221 BASED ON        1 DEGREES OF FREEDOM
PROBABILITY VALUE FOR THE CHI-SQUARE STATISTIC IS       0.63826
THE NORMAL THEORY RLS CHI-SQUARE FOR THIS ML SOLUTION IS           0.221.

BENTLER-BONETT NORMED     FIT INDEX=       1.000
BENTLER-BONETT NONNORMED FIT INDEX=        1.000
COMPARATIVE FIT INDEX            =         1.000
```

```
                    ITERATIVE SUMMARY\

                    PARAMETER
ITERATION           ABS CHANGE          ALPHA              FUNCTION
    1               14.687761          1.00000             5.15066
    2               14.361547          1.00000             0.00170
    3                0.000366          1.00000             0.00148
```

A saturated model provides a highly useful benchmark against which the fit of any nonsaturated model can be tested. Consider the fact that the model examined here is nested in the saturated model presented in Fig. 9. In fact, the current model differs only in one aspect from the saturated model (i.e., in the imposed restriction of equality on the two path coefficients that pertain to the substantive query), and therefore can be examined by using the chi-square difference test. Subtracting the values of the two chi-squares (i.e., 0.221 − 0 = 0.221) and subtracting the degrees of freedom of the two models (i.e., 1 − 0 = 1) provide a nonsignificant result when judged against the chi-square distribution with 1 degree of freedom. This is because the cut-off value for the chi-square distribution with 1 degree of freedom is 3.84 at a significance level .05 (and obviously higher for any smaller level), as can be found in any table of chi-square critical values. Thus, the conclusion is that there is no evidence in the data to warrant rejection of the introduced restriction. That is, the hypothesis of equality of the impacts of IQ and ED_MOTIV on AV_ELECT can be retained.

```
MEASUREMENT EQUATIONS WITH STANDARD ERRORS AND TEST STATISTICS

AV_REQRD=V1   =     .085*V3    +   .006*V4   +  -.012*V5    + 1.000 E1
                    .007             .013          .024
                  12.109             .508         -.488

AV_ELECT=V2   =     .045*V3    +   .144*V4   +   .144*V5    + 1.000 E2
                    .006             .012          .012
                   7.057           12.337        12.337
```

As requested in the input file, the solution equalizes the impact of IQ (V4) and ED_MOTIV (V5) on AV_ELECT (V2). Indeed, both pertinent path coefficients are estimated at .144 and are significant. Note that all other

path coefficients are significant except the ones relating IQ and
ED_MOTIV to AV_REQRD (V1). This issue is considered further at the end
of the discussion of the output.

```
VARIANCES OF INDEPENDENT VARIABLES
----------------------------------
```

	V		F
	---		---
V3 - SAT	47.457*I		I
	5.498 I		I
	8.631 I		I
	I		I
V4 - IQ	10.267*I		I
	1.190 I		I
	8.631 I		I
	I		I
V5 -ED_MOTIV	2.675*I		I
	.310 I		I
	8.631 I		I
	I		I

```
VARIANCES OF INDEPENDENT VARIABLES
----------------------------------
```

	E		D
	---		---
E1 -AV_REQRD	.257*I		I
	.030 I		I
	8.631 I		I
	I		I
E2 -AV_ELECT	.236*I		I
	.027 I		I
	8.631 I		I
	I		I

```
COVARIANCES AMONG INDEPENDENT VARIABLES
---------------------------------------
```

	V		F
	---		---
V4 - IQ	4.100*I		I
V3 - SAT	1.839 I		I
	2.229 I		I
	I		I
V5 -ED_MOTIV	6.394*I		I
V3 - SAT	1.061 I		I
	6.025 I		I
	I		I
V5 -ED_MOTIV	.525*I		I
V4 - IQ	.431 I		I
	1.217 I		I
	I		I

```
COVARIANCES AMONG INDEPENDENT VARIABLES
-----------------------------------------

                          E                        D
                         ---                      ---
E2 -AV_ELECT            .171*I                      I
E1 -AV_REQRD            .025 I                      I
                      6.974 I                      I
                           I                      I
```

As can also be seen from the model in the preceding section, the covariance between IQ and ED_MOTIV is not significant. This suggests that there is weak evidence in the analyzed data set for a (linear) relationship between these two high school variables.

```
STANDARDIZED SOLUTION:

AV_REQRD=V1  =    .761*V3    +   .027*V4    + -.025*V5    +   .658 E1
AV_ELECT=V2  =    .354*V3    +   .530*V4    +  .271*V5    +   .559 E2
```

It is important to note that the standardized solution has not upheld the introduced restriction. Indeed, the standardized path coefficients of V4 and V5 on V2 are not equal here, unlike in the (immediately preceding) unstandardized solution, which fulfilled this restriction. This violation of constraints in the standardized solution is a common finding with restricted models partly because typically the constraints are imposed on parameters that do not have identical units of measurement to begin with. Thus, standardization of the variable metrics, which underlies the standardized solution, destroys the restrictions because it affects differentially the original units of assessment in various variables. Nonetheless, the restriction imposed in the model fitted in this section has an effect and leads to a standardized solution distinct from the one for the unrestricted model presented earlier, as shown in this output portion. The reason for this is that the standardization is carried out on the solution obtained with the constrained model, by subsequently modifying its parameter estimates appropriately.

```
CORRELATIONS AMONG INDEPENDENT VARIABLES
-----------------------------------------

                          V                        F
                         ---                      ---
V4 - IQ                 .186*I                      I
V3 - SAT                     I                      I
                             I                      I
V5 -ED_MOTIV            .567*I                      I
V3 - SAT                     I                      I
                             I                      I
V5 -ED_MOTIV            .100*I                      I
V4 - IQ                      I                      I
                             I                      I
```

```
CORRELATIONS AMONG INDEPENDENT VARIABLES
------------------------------------------
                        E                         D
                       ---                       ---
E2 -AV_ELECT                    .696*I                        I
E1 -AV_REQRD                         I                        I
                                     I                        I
------------------------------------------------------------------------
                    E N D    O F    M E T H O D
------------------------------------------------------------------------
```

As with the model presented in Fig. 9, the last output portion shows that there is a substantial degree of interrelationship between the error terms. That is, those parts of AV_REQRD and AV_ELECT that are unexplained (linearly) by the common set of predictors used are markedly related. This finding again provides some hint concerning the possibility of common omitted variables, as mentioned at the end of the discussion of the unconstrained model results.

MODEL MODIFICATIONS

The restricted model examined in the previous section indicated that the path coefficients relating IQ and ED_MOTIV to AV_REQRD were nonsignificant. An interesting question to consider is whether one or both of these paths can be excluded in a modified model and a good model fit can still be obtained for the data description. As indicated in chapter 1, the modification of a specified model has been termed a specification search (Long, 1983). As such, a specification search is conducted with the intent to detect and correct specification errors between a proposed model and the true model characterizing the population and variables in the study. Although some new search procedures have been developed to automate the process of model modification using computer algorithms, in this introductory book only user-specified model modifications are discussed (for more extensive discussions, see Marcoulides et al., 1998; Marcoulides & Drezner, in press; Scheines, Spirtes, & Glymour, in press; Scheines, Spirtes, Glymour, Meek, & Richardson, 1998).

Consider a modification of the model in which the path coefficients relating IQ and ED_MOTIV to AV_REQRD were found to be nonsignificant. Because it is well known that a single change in a model can affect other parts of the solution, only one change at a time is made in the model. By proceeding step by step (i.e., by setting one parameter to 0 at a time) the possibility of missing a tenable restrictive model or misspecifying a model, which might occur if all the nonsignificant parameters were fixed to 0 at once, are excluded. In a sense, this strategy is analogous to backward selection in regression analysis, particularly the rule not to drop from the equation more than a single predictor during a step (although automated model-selection search procedures have also recently been developed for regression analyses; see Drezner, Marcoulides, & Salhi, 1999; Marcoulides & Drezner, 1999).

With user-specified (nonautomated) model specification searches there are no rules concerning which nonsignificant parameters to fix to 0 first. One strategy might be to consider setting to 0 the parameter with the smallest t value taken as an absolute value, which in a sense might appear most nonsignificant. In the example, this is the path coefficient of ED_MOTIV on AV_REQRD. Fixing this path coefficient to 0 yields a nested model because the resulting model is obtained from that in Fig. 9 after introducing a parameter restriction. The fixing is achieved by simply not mentioning the constrained parameter as free when defining the model or, alternatively, by fixing it after the MOdel definition line. Provided here is the LISREL input file for this model with the restricted parameter between ED_MOTIV and AV_REQRD, GA (1, 3).

```
PATH ANALYSIS MODEL FOR RESTRICTED MODEL WITH THE PATH
OF ED_MOTIV TO AV_REQRD FIXED AT ZERO
DA NI=5 NO=150
CM
.594
.483    .754
3.993   3.626   47.457
.426    1.757   4.100    10.267
.500    .722    6.394    .525     2.675
LA
AV_REQRD AV_ELECT SAT IQ ED_MOTIV
MO NY=2 NX=3 GA=FU,FR PH=SY,FR PS=SY,FR
EQ GA(2, 2) GA(2, 3)
FI GA(1, 3) ! ADDED LINE IN ORDER TO FIX THE PARAMETER γ₁₃
OU
```

(From here onward, to save space, we adopt the practice of reporting only those parts of the output, or fit indices if pertinent, that are relevant for answering the substantive query that has led to the fitted model.)

The model for the LISREL input file just presented yields a chi-square value $T = 0.46$ for 2 degrees of freedom. Thus, the difference in chi-square values between this model and the immediately preceding one with only the constraint of two equal path coefficients, is $\Delta T = 0.24$ for $\Delta df = 1$. Because this difference is nonsignificant (when compared to the cutoff of 3.84 for the chi-square distribution with 1 degree of freedom), one can conclude that there is no evidence in our data concerning the influence of ED_MOTIV on AV_REQRD, after accounting for the impact of SAT and IQ (note that these two predictors are still present in the equation for this dependent variable).

Because the imposed restriction was found to be plausible, it is retained in the model and the next nonsignificant path (i.e., the one between IQ and AV_REQRD) is considered. Fixing this path coeffient to 0 (by including the statement FI GA(1, 2) in the LISREL input file) provides a resulting chi-square value of $T = 0.87$, for $df = 3$. Thus, the increase in chi-square is $\Delta T = 0.41$ for $\Delta df = 1$ and is similarly nonsignificant. One can therefore conclude that there is no evidence in the data concerning the impact of IQ on AV_REQRD, once that of SAT is accounted for.

Following these modifications, the model does not have any more nonsignificant parameters and represents the most restrictive, yet tenable model examined in this chapter. The results following this specification search are presented here (with some redundant sections eliminated):

LISREL ESTIMATES (MAXIMUM LIKELIHOOD)

GAMMA

	SAT	IQ	ED_MOTIV
AV_REQRD	.08	- -	- -
	(.01)		
	13.79		
AV_ELECT	.05	.14	.14
	(.01)	(.01)	(.01)
	7.37	16.72	16.72

COVARIANCE MATRIX OF Y AND X

	AV_REQRD	AV_ELECT	SAT	IQ	ED_MOTIV
AV_REQRD	.59				
AV_ELECT	.48	.75			
SAT	3.99	3.63	47.46		
IQ	.34	1.71	4.10	10.27	
ED_MOTIV	.54	.74	6.39	.53	2.68

PHI

	SAT	IQ	ED_MOTIV
SAT	47.46		
IQ	4.10	10.27	
ED_MOTIV	6.39	.53	2.68

PSI

	AV_REQRD	AV_ELECT
AV_REQRD	.26	
	(.03)	
	8.54	
AV_ELECT	.17	.24
	(.02)	(.03)
	6.90	8.54

SQUARED MULTIPLE CORRELATIONS FOR STRUCTURAL EQUATIONS

AV_REQRD	AV_ELECT
.57	.68

Although this final model does not perfectly reproduce the analyzed covariance matrix (indeed, in its section entitled COVARIANCE MATRIX OF Y AND X quite a few of its entries marginally deviate from their counterparts in the sample covariance matrix), the fit criteria indicate that it is a plausible model. Despite these good fit criteria, it is important to note that results obtained from any model specification search may be unique to the particular data set and that chance elements may affect the search. In fact, once a specification search is conducted, a researcher is actually entering a more exploratory phase of analysis. Hence, the possibility exists that the results regarding aspects of the model are due only to chance fluctuations. For this reason, any models that result from specification searches should always be cross-validated before any real validity can be claimed.

CHAPTER FOUR

Confirmatory Factor Analysis

WHAT IS FACTOR ANALYSIS?

Factor analysis is an approach that was first developed by psychologists as a way to represent latent (hypothetically existing) variables. Although the latent variables could not be directly measured, psychologists still wanted to handle them as if they were measurable. For example, socioeconomic status is commonly discussed as if it is a real measured variable. Nevertheless, it is a latent variable that is generally regarded as a way to describe an individual's condition of wealth using indicators such as occupation, type of car owned, and place of residence.

Factor analysis has a relatively long history. The idea goes back to the early 1900s and it is generally acknowledged that the Englishman Charles Spearman first applied the approach to study the structure of human abilities. Spearman proposed that an individual's ability scores were simply manifestations of some general ability (called general intelligence, or just g) and other specific abilities (such as verbal or numerical abilities). The general and specific factors combined to produce the ability performance. This idea was labeled the *two-factor theory of human abilities*. However, as more researchers became interested in this approach (e.g., Thurstone, 1935), the theory was extended to models with more factors and was referred to as the *common-factor theory* or just factor analysis.

In general terms, *factor analysis* is an approach for expressing in the language of mathematics hypothetical constructs by using a variety of observable indicators that can be directly measured. The analysis is considered exploratory when the concern is with determining how many constructs (factors) are needed to explain the relationships among the observed indicators and confirmatory when a preexisting model of the relationship among the indicators directs the search. As such, confirmatory factor analysis (CFA) is not concerned with discovering a factor structure, but with confirming the existence of a specific factor structure. Of course, in order to confirm the existence of a specific factor structure one must have some initial idea about the composition of the structure. In this respect, confirmatory factor analysis is considered to be a general modeling approach that is designed to test hypotheses about a factor structure whose number and interpretation are given in advance. Hence, in confirmatory factor analysis (a) the theory comes first, (b) the model is then derived from it, and finally (c) the model is tested for consistency with the observed data using a SEM-type approach. Therefore, as discussed at length in chapter 1, the unknown model parameters are chosen so that, in general, the model-reproduced Σ matrix comes as close as possible to the sample matrix S (thus, in a sense, the model is given the best chance to emulate S). If the proposed model emulates S to a sufficient extent (as measured by the goodness-of-fit indices), it can be treated as a plausible description of the phenomenon under investigation and the theory from which the model has been derived is supported. Otherwise, the model is rejected and the theory—as embodied in the model—is disconfirmed. It is important to note that this testing rationale is valid for all applications of the SEM methodology, not only those within the framework of confirmatory factor analysis, but its origins are firmly rooted in the factor analytic approach.

This discussion of confirmatory factor analysis (CFA) suggests an important limitation concerning its use. The starting point of CFA is a very demanding one, requiring that the complete details of a proposed model be specified before being fitted to the data. Unfortunately, in many substantive areas this may be too strong a requirement because theories are often poorly developed or even nonexistent. Because of these potential limitations, Jöreskog & Sörbom (1993a) distinguished three situations concerning model fitting and testing: (a) the strictly confirmatory situation in which a single formulated model is either accepted or rejected, (b) the alternative-models or competing-models situation in which several models are formulated and one of them is selected, and (c) the model-generating situation in which an initial model is specified and, if it does not fit the data, is modified and repeatedly tested until some fit is obtained.

It should be obvious that the strictly confirmatory situation is rare in practice because most researchers are simply not willing to reject a proposed model without at least suggesting some alternative model. The alternative- or competing-model situation is also not very common because researchers usually prefer not to specify alternative models beforehand.

Thus, model generating is the most common situation encountered in practice (Jöreskog & Sörbom, 1993a; Marcoulides, 1989). As a consequence, many applications of CFA actually bear some characteristic features of both explanatory and confirmatory approaches. As such, it is not very frequent that researchers are dealing with purely exploratory or purely confirmatory analyses. It is for this reason that the results of any repeated analyses of models on the same data set should be treated with a great deal of caution and be considered tentative until a replication study can provide further information on the performance of the generated models.

AN EXAMPLE CONFIRMATORY
FACTOR ANALYSIS MODEL

To demonstrate a confirmatory factor analysis model, consider the following example model with three latent variables: ability, achievement motivation, and aspiration. The model proposes that three variables are indicators of ability, three variables are indicators of achievement motivation, and two variables are indicators of aspiration. Here the primary interest is in estimating the relationships among ability, motivation, and aspiration. For the purposes of this study, data were collected from a sample of $N = 250$ second-year college students. The following observed variables were used in the study:

1. A general ability score (ABILITY1).
2. Grade point average obtained in last year of high school (ABILITY2).
3. Grade point average obtained in first year of college (ABILITY3).
4. Achievement motivation score 1 (MOTIVN1).
5. Achievement motivation score 2 (MOTIVN2).
6. Achievement motivation score 3 (MOTIVN3).
7. A general educational aspiration score (ASPIRN1).
8. A general vocational aspiration score (ASPIRN2).

The example confirmatory factor analysis model is presented in Fig. 10 and the observed covariance matrix is presented in Table 1. The model is initially presented in EQS notation using V_1 to V_8 for the observed variables, E_1 to E_8 for the error terms associated with the observed variables, and F_1 to F_3 for the latent variables.

To determine the parameters of the model presented in Fig. 10 (designated by asterisks), one must follow the six rules outlined in chapter 1. According to Rule 1, all eight error-term variances are model parameters

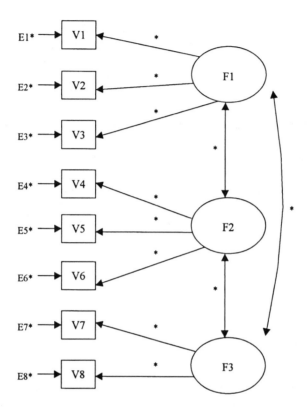

FIG. 10.   Example confirmatory factor analysis model using EQS notation.

TABLE 1
Covariance Matrix for Confirmatory Factor Analysis
Example of Ability, Motivation, and Aspiration

Variable	AB1	AB2	AB3	MOT1	MOT2	MOT3	ASP1	ASP2
AB1	.45							
AB2	.32	.56						
AB3	.27	.32	.45					
MOT1	.17	.20	.19	.55				
MOT2	.20	.21	.18	.30	.66			
MOT3	.19	.25	.20	.30	.36	.61		
ASP1	.08	.12	.09	.23	.27	.22	.58	
ASP2	.11	.10	.07	.21	.25	.27	.39	.62

Notes.   AB denotes ability; MOT, motivation; ASP, aspiration. Sample size = 250.

97

and, according to Rule 3, the eight factor loadings are also model parameters. In addition, the three construct variances are tentatively designated model parameters (but see the use of Rule 6 later in the paragraph). Following Rule 2, the three covariances between latent variables are also model parameters. Rule 4 is not applicable to this model because no explanatory relationships are assumed among any of the latent variables. For Rule 5, observe that there are no two-way arrows connecting dependent variables, or a dependent and independent variable in Fig. 10. Finally, Rule 6 requires that the scale of each latent variable be fixed. Because this study's primary interest is in estimating the correlations between ability and motivation (which are identical to their covariances if the variances of the latent variables are set equal to 1), the variances of the latent variables are simply fixed to unity. This decision makes the construct variances fixed parameters rather than free model parameters (as originally designated). Thus, the model in Fig. 10 has, altogether, 19 parameters ($8 + 3 + 8 = 19$), symbolized by asterisks.

EQS AND LISREL INPUT FILES

EQS Input File

The EQS input file is constructed following the principles outlined in chapter 2. Accordingly, the file begins with a title command line followed by a specification line providing the number of variables in the model and sample size.

```
/TITLE
EXAMPLE CONFIRMATORY FACTOR ANALYSIS;
/SPECIFICATIONS
CASES=250; VARIABLES=8;
```

To facilitate interpretation of the output, labels are provided for all variables included in the model using the command line /LABELS.

```
/LABELS
V1=ABILITY1; V2=ABILITY2; V3=ABILITY3; V4=MOTIVN1;
V5=MOTIVN2; V6=MOTIVN3; V7=ASPIRN1; V8=ASPIRN2;
F1=ABILITY; F2=MOTIVATN; F3=ASPIRATN;
```

Next the model definition equations are provided followed by the remaining model parameters in the variance and covariance command lines.

```
/EQUATIONS
V1=*F1+E1;
V2=*F1+E2;
V3=*F1+E3;
V4=*F2+E4;
V5=*F2+E5;
V6=*F2+E6;
V7=*F3+E7;
V8=*F3+E8;
/VARIANCES
F1 TO F3=1; E1 TO E8=*;
/COVARIANCES
F1 TO F3=*;
```

Finally, the data are provided along with the end of the input file command line.

```
/MATRIX
.45
.32   .56
.27   .32   .45
.17   .20   .19   .55
.20   .21   .18   .30   .66
.19   .25   .20   .30   .36   .61
.08   .12   .09   .23   .27   .22   .58
.11   .10   .07   .21   .25   .27   .39   .62
/END;
```

The complete EQS input file is as follows (using the appropriate abbreviations):

```
/TIT
EXAMPLE CONFIRMATORY FACTOR ANALYSIS;
/SPE
CAS=250; VAR=8;
/LAB
V1=ABILITY1; V2=ABILITY2; V3=ABILITY3; V4=MOTIVN1;
V5=MOTIVN2; V6=MOTIVN3; V7=ASPIRN1; V8=ASPIRN2;
F1=ABILITY; F2=MOTIVATN; F3=ASPIRATN;
/EQU
V1=*F1+E1;
V2=*F1+E2;
V3=*F1+E3;
```

```
V4=*F2+E4;
V5=*F2+E5;
V6=*F2+E6;
V7=*F3+E7;
V8=*F3+E8;
/VAR
F1 TO F3=1; E1 TO E8=*;
/COV
F1 TO F3=*;
/MAT
.45
.32  .56
.27  .32  .45
.17  .20  .19  .55
.20  .21  .18  .30  .66
.19  .25  .20  .30  .36  .61
.08  .12  .09  .23  .27  .22  .58
.11  .10  .07  .21  .25  .27  .39  .62
/END;
```

LISREL Input File

The corresponding LISREL input file is presented next. The results for
both formats are provided in a separate section. The LISREL input file is
constructed using the principles outlined in chapter 2.

```
EXAMPLE CONFIRMATORY FACTOR ANALYSIS
DA NI=8 NO=250
CM
.45
.32  .56
.27  .32  .45
.17  .20  .19  .55
.20  .21  .18  .30  .66
.19  .25  .20  .30  .36  .61
.08  .12  .09  .23  .27  .22  .58
.11  .10  .07  .21  .25  .27  .39  .62
LA
ABILITY1 ABILITY2 ABILITY3 MOTIVN1 MOTIVN2 MOTIVN3 C
ASPIRN1 ASPIRN2
MO NY=8 NE=3 PS=SY,FR TE=DI,FR LY=FU,FI
LE
ABILITY MOTIVATN ASPIRATN
```

```
FR LY(1, 1) LY(2, 1) LY(3, 1)
FR LY(4, 2) LY(5, 2) LY(6, 2)
FR LY(7, 3) LY(8, 3)
FI PS(1, 1) PS(2, 2) PS(3, 3)
VA 1 PS(1, 1) PS(2, 2) PS(3, 3)
OU
```

Using the same title, the data definition line declares that the model will be fit to data on eight variables collected from 250 subjects. The sample covariance matrix CM is provided next, along with the variable labels (note the use of C, for Continue, to wrap over to the second label line). The latent variables are also provided with labels by using the notation LE (for Labels for Etas; following the notation of the general LISREL model in which the Greek letter *eta* represents a latent variable). Finally, in the model command line, the three matrices PS, TE, and LY are declared. The latent variances and covariance matrix PS is initially declared to be symmetric and free (i.e., all its elements are declared to be free parameters), which defines all factor variances and covariances as model parameters. Subsequently, for the reasons discussed in the previous section, the variances in the matrix PS are fixed to a value of 1. The error covariance matrix TE is defined as diagonal (i.e., no error covariances are introduced) and thus only has as model parameters all the error variances along its main diagonal. Defining, then, the matrix of factor loadings LY as a fixed and full (rectangular) matrix relating the eight manifest variables to the three latent variables permits those loadings that relate the corresponding indicators to their factors to be declared freed in the next lines.

MODELING RESULTS

EQS Program Results

The EQS input described in the previous section produces the following results. In presenting the output section of the EQS program, comments are inserted at appropriate places. The pages echoing the input and the recurring page titles are omitted in order to save space.

```
MAXIMUM LIKELIHOOD SOLUTION (NORMAL DISTRIBUTION THEORY)

PARAMETER ESTIMATES APPEAR IN ORDER,
NO SPECIAL PROBLEMS WERE ENCOUNTERED DURING OPTIMIZATION.
```

This message indicates that the program has not encountered problems stemming from lack of model identification or other numerical difficulties and is a reassuring message that the model is technically sound.

```
RESIDUAL COVARIANCE MATRIX (S-SIGMA):

                  ABILITY1    ABILITY2    ABILITY3    MOTIVTN1    MOTIVN2
                   V   1       V   2       V   3       V   4       V   5
ABILITY1 V   1     0.000
ABILITY2 V   2     0.001       0.000
ABILITY3 V   3    -0.001      -0.001       0.000
MOTIVTN1 V   4     0.001       0.000       0.020       0.000
MOTIVN2  V   5     0.004      -0.022      -0.017      -0.004       0.000
MOTIVN3  V   6    -0.007       0.017       0.002      -0.004       0.007
ASPIRTN1 V   7    -0.008       0.016       0.001       0.016       0.022
ASPIRTN2 V   8     0.019      -0.008      -0.022      -0.011      -0.007


                  MOTIVN3     ASPIRTN1    ASPIRTN2
                   V   6       V   7       V   8
MOTIVN3  V   6     0.000
ASPIRTN1 V   7    -0.029       0.000
ASPIRTN2 V   8     0.013       0.000       0.000


            AVERAGE ABSOLUTE COVARIANCE RESIDUALS        =        0.0077
       AVERAGE OFF-DIAGONAL ABSOLUTE COVARIANCE RESIDUALS   =       0.0099


STANDARDIZED RESIDUAL MATRIX:

                  ABILITY1    ABILITY2    ABILITY3    MOTIVTN1    MOTIVN2
                   V   1       V   2       V   3       V   4       V   5
ABILITY1 V   1     0.000
ABILITY2 V   2     0.002       0.000
ABILITY3 V   3    -0.001      -0.001       0.000
MOTIVTN1 V   4     0.002       0.000       0.041       0.000
MOTIVN2  V   5     0.007      -0.037      -0.031      -0.006       0.000
MOTIVN3  V   6    -0.012       0.029       0.005      -0.008       0.011
ASPIRTN1 V   7    -0.016       0.027       0.003       0.028       0.035
ASPIRTN2 V   8     0.036      -0.013      -0.041      -0.019      -0.010

                  MOTIVN3     ASPIRTN1    ASPIRTN2
                   V   6       V   7       V   8
MOTIVN3  V   6     0.000
ASPIRTN1 V   7    -0.049       0.000
ASPIRTN2 V   8     0.021       0.000       0.000


            AVERAGE ABSOLUTE STANDARDIZED RESIDUALS       =        0.0137
       AVERAGE OFF-DIAGONAL ABSOLUTE STANDARDIZED RESIDUALS   =      0.0176


LARGEST STANDARDIZED RESIDUALS:

   V  7,V  6     V  8,V  3     V  4,V  3     V  5,V  2     V  8,V  1
     -0.049        -0.041        0.041        -0.037        0.036

   V  7,V  5     V  5,V  3     V  6,V  2     V  7,V  4     V  7,V  2
      0.035        -0.031        0.029         0.028         0.027
```

```
V  8,V  6      V  8,V  4      V  7,V  1      V  8,V  2      V  6,V  1
   0.021         -0.019         -0.016         -0.013         -0.012

V  6,V  5      V  8,V  5      V  6,V  4      V  5,V  1      V  5,V  4
   0.011         -0.010         -0.008          0.007         -0.006
```

DISTRIBUTION OF STANDARDIZED RESIDUALS

```
       - - - - - - - - - - - - - - - - - - - - - - - - - -
       !                                    !
  40 -                                      -
       !                                    !
       !                                    !
       !                                    !
       !                                    !       RANGE      FREQ  PERCENT
  30 -                                      -
       !                                    !  1  -0.5  -  --       0    0.00%
       !                                    !  2  -0.4  - -0.5      0    0.00%
       !                                    !  3  -0.3  - -0.4      0    0.00%
       !                    *               !  4  -0.2  - -0.3      0    0.00%
  20 -                      *               -  5  -0.1  - -0.2      0    0.00%
       !                    *               !  6   0.0  - -0.1     22   61.11%
       !                    *               !  7   0.1  -  0.0     14   38.89%
       !                    *  *            !  8   0.2  -  0.1      0    0.00%
       !                    *  *            !  9   0.3  -  0.2      0    0.00%
  10 -                      *  *            -  A   0.4  -  0.3      0    0.00%
       !                    *  *            !  B   0.5  -  0.4      0    0.00%
       !                    *  *            !  C   ++   -  0.5      0    0.00%
       !                    *  *            !     - - - - - - - - - - - - - - - - -
       !                    *  *            !         TOTAL        36  100.00%
       - - - - - - - - - - - - - - - - - - - - - -
```

EACH "*" REPRESENTS 2 RESIDUALS

None of the residuals presented in this section of the output are a cause
for concern. This is typically the case for well-fitting models with regard to
all variable variances and covariances. Note the effectively symmetric
shape of their distribution (observe the range of variability of their magnitude on the right-hand side of this output portion).

```
GOODNESS OF FIT SUMMARY
INDEPENDENCE MODEL CHI-SQUARE =        801.059  ON     28 DEGREES OF FREEDOM
INDEPENDENCE AIC =    745.05881    INDEPENDENCE CAIC =    618.45790
        MODEL AIC =    -13.41865         MODEL CAIC =     -90.28349

CHI-SQUARE =      20.581 BASED ON     17 DEGREES OF FREEDOM
PROBABILITY VALUE FOR THE CHI-SQUARE STATISTIC IS       0.24558
THE NORMAL THEORY RLS CHI-SQUARE FOR THIS ML SOLUTION IS          18.891.

BENTLER-BONETT NORMED     FIT INDEX=       0.974
BENTLER-BONETT NONNORMED FIT INDEX=        0.992
COMPARATIVE FIT INDEX            =         0.995
```

```
                        ITERATIVE SUMMARY

                   PARAMETER
    ITERATION       ABS CHANGE        ALPHA          FUNCTION
        1            0.308231        1.00000         0.54912
        2            0.141164        1.00000         0.24514
        3            0.061476        1.00000         0.10978
        4            0.013836        1.00000         0.08281
        5            0.001599        1.00000         0.08266
        6            0.000328        1.00000         0.08266
```

The goodness-of-fit indices are satisfactory and identical (within round-off error) to those provided by LISREL later. Note in particular that the Bentler–Bonett indices, as well as the comparative fit index, are all in the high .90s and suggest a fairly good model fit. The ITERATIVE SUMMARY, which provides an account of the numerical routine performed by EQS to minimize the maximum-likelihood fit function, also indicates a quick and trouble-free convergence to the final solution reported next.

```
MEASUREMENT EQUATIONS WITH STANDARD ERRORS AND TEST STATISTICS

    ABILITY1=V1    =         .519*F1    + 1.000 E1
                             .039
                           13.372

    ABILITY2=V2    =         .615*F1    + 1.000 E2
                             .043
                           14.465

    ABILITY3=V3    =         .521*F1    + 1.000 E3
                             .039
                           13.461

    MOTIVTN1=V4    =         .511*F2    + 1.000 E4
                             .045
                           11.338

    MOTIVN2 =V5    =         .594*F2    + 1.000 E5
                             .049
                           12.208

    MOTIVN3 =V6    =         .595*F2    + 1.000 E6
                             .046
                           12.882

    ASPIRTN1=V7    =         .614*F3    + 1.000 E7
                             .049
                           12.551

    ASPIRTN2=V8    =         .635*F3    + 1.000 E8
                             .051
                           12.547
```

This is the final model solution presented in nearly the same form as the model equations provided to EQS in the input file (recall that in EQS, asterisks denote the estimated parameters). Immediately beneath each parameter estimate its standard error appears and below the standard errors the *t* values. As can also be observed in the LISREL output presented later, the equations generated by EQS suggest that some of the indicators load to a very similar degree on their factors (this issue is revisited in a later section of the chapter when some restricted hypotheses are tested).

```
VARIANCES OF INDEPENDENT VARIABLES
------------------------------

            V                               F
            ---                             ---
                  I F1    -ABILITY              1.000 I
                  I                                   I
                  I                                   I
                  I                                   I
                  I F2    -MOTIVATN              1.000 I
                  I                                   I
                  I                                   I
                  I                                   I
                  I F3    -ASPIRATN              1.000 I
                  I                                   I
                  I                                   I
                  I                                   I
```

These are the variances of the latent variables set to a value of 1.

```
VARIANCES OF INDEPENDENT VARIABLES
------------------------------

              E                         D
              ---                       ---
E1 -ABILITY1          .181*I                  I
                      .023 I                  I
                     7.936 I                  I
                           I                  I
E2 -ABILITY2          .182*I                  I
                      .027 I                  I
                     6.673 I                  I
                           I                  I
E3 -ABILITY3          .178*I                  I
                      .023 I                  I
                     7.846 I                  I
                           I                  I
E4 -MOTIVTN1          .288*I                  I
                      .032 I                  I
                     8.965 I                  I
                           I                  I
```

```
E5  -MOTIVN2                    .308*I                          I
                               .037 I                          I
                              8.362 I                          I
                                    I                          I
E6  -MOTIVN3                    .256*I                          I
                               .033 I                          I
                              7.760 I                          I
                                    I                          I
E7  -ASPIRTN1                   .203*I                          I
                               .040 I                          I
                              5.109 I                          I
                                    I                          I
E8  -ASPIRTN2                   .217*I                          I
                               .042 I                          I
                              5.118 I                          I
                                    I                          I
```

These are the variances of the residual terms along with their standard errors and *t* values.

```
COVARIANCES AMONG INDEPENDENT VARIABLES
-------------------------------------
                 V                              F
                ---                            ---
                        I F2    -MOTIVATN              .636*I
                        I F1    -ABILITY               .055 I
                        I                            11.678 I
                        I                                   I
                        I F3    -ASPIRATN              .276*I
                        I F1    -ABILITY               .073 I
                        I                             3.764 I
                        I                                   I
                        I F3    -ASPIRATN              .681*I
                        I F2    -MOTIVATN              .054 I
                        I                            12.579 I
```

These are the intercorrelations among the latent variables along with their standard errors and *t* values (note that within rounding-off errors they are essentially identical to the ones obtained with LISREL later).

```
STANDARDIZED SOLUTION:

ABILITY1=V1   =    .773*F1   + .634 E1
ABILITY2=V2   =    .822*F1   + .570 E2
ABILITY3=V3   =    .777*F1   + .629 E3
MOTIVTN1=V4   =    .690*F2   + .724 E4
MOTIVN2=V5    =    .731*F2   + .683 E5
MOTIVN3=V6    =    .762*F2   + .648 E6
ASPIRTN1=V7   =    .807*F3   + .591 E7
ASPIRTN2=V8   =    .806*F3   + .592 E8
```

As discussed in chapter 3, this STANDARDIZED SOLUTION output results from standardizing all the variables in the model. Because the standardized solution uses a metric that is uniform across all measures, it is possible to address the issue of the relative importance of the manifest variables in assessing the underlying constructs by comparing their loadings.

```
CORRELATIONS AMONG INDEPENDENT VARIABLES
-----------------------------------------

              V                                    F
             ---                                  ---
                    I F2   -MOTIVATN              .636*I
                    I F1   -ABILITY                    I
                    I                                   I
                    I F3   -ASPIRATN              .276*I
                    I F1   -ABILITY                    I
                    I                                   I
                    I F3   -ASPIRATN              .681*I
                    I F2   -MOTIVATN                   I
                    I                                   I
```

These are the correlations among the latent variables included in the confirmatory factor analysis model. The output section leads one to infer that there are significant and medium-size relationships between ability and motivation (estimated at .64), as well as between motivation and aspiration (estimated at .68). In contrast, the correlation between ability and aspiration appears to be much weaker (estimated at .28).

LISREL Program Results

The output produced by the LISREL input file created in the previous section is presented next. Comments are inserted at appropriate places to clarify portions of the output and, for brevity, the echoed input file, the analyzed covariance matrix, and the recurring first title line are not presented.

```
PARAMETER SPECIFICATIONS

        LAMBDA-Y

          ABILITY   MOTIVATN   ASPIRATN
          --------  --------   --------
ABILITY1      1         0          0
ABILITY2      2         0          0
ABILITY3      3         0          0
MOTIVTN1      0         4          0
MOTIVTN2      0         5          0
MOTIVTN3      0         6          0
ASPIRTN1      0         0          7
ASPIRATN      0         0          8
```

```
        PSI

              ABILITY   MOTIVATN   ASPIRATN
              --------  --------   --------
ABILITY           0
MOTIVATN          9         0
ASPIRATN         10        11          0

        THETA-EPS

              ABILITY1  ABILITY2  ABILITY3  MOTIVTN1  MOTIVTN2  MOTIVTN3
              --------  --------  --------  --------  --------  --------
                 12        13        14        15        16        17

        THETA-EPS

              ASPIRTN1  ASPIRATN
              --------  --------
                 18        19
```

Observe from this section that the input file has correctly communicated to the LISREL program the number and exact location of the 19 model parameters—eight factor loadings, three factor variances, and eight error variances.

```
LISREL ESTIMATES (MAXIMUM LIKELIHOOD)

        LAMBDA-Y

              ABILITY   MOTIVATN   ASPIRATN
              --------  --------   --------
ABILITY1         .52        - -         - -
                (.04)
                13.37

ABILITY2         .61        - -         - -
                (.04)
                14.46

ABILITY3         .52        - -         - -
                (.04)
                13.46

MOTIVTN1         - -        .51         - -
                          (.05)
                          11.34

MOTIVTN2         - -        .59         - -
                          (.05)
                          12.21

MOTIVTN3         - -        .60         - -
                          (.05)
                          12.89

ASPIRTN1         - -        - -         .61
                                      (.05)
                                      12.55
```

```
ASPIRATN           - -         - -          .63
                                          (.05)
                                          12.55
```

This section of the output presents the factor loading estimates of the LY matrix along with their standard errors and *t* values in a column format (note that within rounding-off errors they are essentially identical to those obtained with EQS). It is important to note that the estimates of the loadings for each indicator appear quite similar within a factor (as in the EQS output). This will have a bearing on the subsequent formal tests of the tau-equivalence hypotheses (i.e., an assumption that the indicators measure the same latent variable in the same units of measurement) concerning the indicators of ability, motivation, and aspiration.

```
          COVARIANCE MATRIX OF ETA

              ABILITY   MOTIVATN   ASPIRATN
              -------   --------   --------
ABILITY         1.00
MOTIVATN         .64      1.00
ASPIRATN         .28       .68       1.00

          PSI

              ABILITY   MOTIVATN   ASPIRATN
              -------   --------   --------
ABILITY         1.00

MOTIVATN         .64      1.00
               (.05)
               11.68

ASPIRATN         .28       .68       1.00
               (.07)     (.05)
                3.76     12.58
```

Based on this output section, it can be inferred that there are significant and medium-size relationships between ability and motivation (estimated at .64), as well as between motivation and aspiration (estimated at .68). It is important to note that because the latent variances were fixed to 1, the COVARIANCE MATRIX OF ETA is identical to the PSI matrix. Nevertheless, differences between the two matrices will emerge in chapter 5, where there will be explanatory relationships postulated between some of the constructs. Using the confidence intervals (CI) of latent correlations (obtained by adding twice the standard errors to and subtracting twice the standard errors from their parameters), it can be shown that these two correlations are practically indistinguishable in the population. Indeed, their approximate 95% CIs are (.64 ± 2 × .05) = (.54; .74) and (.68 ± 2 ×

.05) = (.58; .78), and overlap to a substantial degree, suggesting this conclusion. In contrast, the correlation between ability and aspiration appears much weaker, although significant; its t value is 3.76, which is well outside the nonsignificance interval (−2; +2). Indeed, this estimated correlation of only .28 explains less than 9% of their interrelationship in the sample.

```
THETA-EPS

          ABILITY1     ABILITY2     ABILITY3     MOTIVTN1     MOTIVTN2     MOTIVTN3
          --------     --------     --------     --------     --------     --------
              .18          .18          .18          .29          .31          .26
            (.02)        (.03)        (.02)        (.03)        (.04)        (.03)
             7.94         6.67         7.85         8.97         8.36         7.76


THETA-EPS

          ASPIRTN1     ASPIRATN
          --------     --------
              .20          .22
            (.04)        (.04)
             5.11         5.12


SQUARED MULTIPLE CORRELATIONS FOR Y - VARIABLES

          ABILITY1     ABILITY2     ABILITY3     MOTIVTN1     MOTIVTN2     MOTIVTN3
          --------     --------     --------     --------     --------     --------
              .60          .67          .60          .48          .53          .58


SQUARED MULTIPLE CORRELATIONS FOR Y - VARIABLES

          ASPIRTN1     ASPIRATN
          --------     --------
              .65          .65
```

Based on this output section, it appears that, apart from the first indicator of motivation (i.e., MOTIVTN1), more than half of the variance in any of the remaining seven measures is explained in terms of latent individual differences on the corresponding factor.

```
                          GOODNESS OF FIT STATISTICS

            CHI-SQUARE WITH 17 DEGREES OF FREEDOM = 20.58 (P = 0.25)
                 ESTIMATED NON-CENTRALITY PARAMETER (NCP) = 3.58
           90 PERCENT CONFIDENCE INTERVAL FOR NCP = (0.0 ; 19.24)

                      MINIMUM FIT FUNCTION VALUE = 0.083
             POPULATION DISCREPANCY FUNCTION VALUE (F0) = 0.014
          90 PERCENT CONFIDENCE INTERVAL FOR F0 = (0.0 ; 0.077)
          ROOT MEAN SQUARE ERROR OF APPROXIMATION (RMSEA) = 0.029
         90 PERCENT CONFIDENCE INTERVAL FOR RMSEA = (0.0 ; 0.067)
            P-VALUE FOR TEST OF CLOSE FIT (RMSEA < 0.05) = 0.78
```

```
           EXPECTED CROSS-VALIDATION INDEX (ECVI) = 0.24
     90 PERCENT CONFIDENCE INTERVAL FOR ECVI = (0.22 ; 0.30)
               ECVI FOR SATURATED MODEL = 0.29
             ECVI FOR INDEPENDENCE MODEL = 3.28

CHI-SQUARE FOR INDEPENDENCE MODEL WITH 28 DEGREES OF FREEDOM = 801.06
               INDEPENDENCE AIC = 817.06
                     MODEL AIC = 58.58
                 SATURATED AIC = 72.00
               INDEPENDENCE CAIC = 853.23
                    MODEL CAIC = 144.49
                SATURATED CAIC = 234.77

          ROOT MEAN SQUARE RESIDUAL (RMR) = 0.011
                 STANDARDIZED RMR = 0.020
           GOODNESS OF FIT INDEX (GFI) = 0.98
     ADJUSTED GOODNESS OF FIT INDEX (AGFI) = 0.96
    PARSIMONY GOODNESS OF FIT INDEX (PGFI) = 0.46

             NORMED FIT INDEX (NFI) = 0.97
           NON-NORMED FIT INDEX (NNFI) = 0.99
       PARSIMONY NORMED FIT INDEX (PNFI) = 0.59
          COMPARATIVE FIT INDEX (CFI) = 1.00
          INCREMENTAL FIT INDEX (IFI) = 1.00
           RELATIVE FIT INDEX (RFI) = 0.96

                  CRITICAL N (CN) = 405.22
```

All of the goodness-of-fit indices indicate an acceptable model. Note in particular the root mean square error of approximation (RMSEA), which is well under the proposed threshold of .05 (see Jöreskog & Sörbom, 1993b; Marcoulides & Hershberger, 1997). Moreover, the left endpoint of its confidence interval, as well as those of the noncentrality parameter and the minimal fit function value in the population, are 0—their best value— which is another sign of a satisfactory model fit.

```
SUMMARY STATISTICS FOR FITTED RESIDUALS
SMALLEST FITTED RESIDUAL =      -.03
   MEDIAN FITTED RESIDUAL =      .00
  LARGEST FITTED RESIDUAL =      .02

STEMLEAF PLOT
 - 2|921
 - 1|71
 - 0|887743110000000000
   0|111247
   1|36679
   2|02

SUMMARY STATISTICS FOR STANDARDIZED RESIDUALS
SMALLEST STANDARDIZED RESIDUAL =    -2.04
```

```
   MEDIAN STANDARDIZED RESIDUAL =      .00
   LARGEST STANDARDIZED RESIDUAL =    1.30

STEMLEAF PLOT
 - 2|0
 - 1|
 - 1|11
 - 0|96
 - 0|444443110000000000
   0|11123
   0|78999
   1|013
```

None of the standardized residuals is very large, suggesting that the model is not only a reasonably good overall means of data description, but also locally performs quite well—no part of the model seems to suffer marked misfit.

```
MODIFICATION INDICES AND EXPECTED CHANGE

        MODIFICATION INDICES FOR LAMBDA-Y
```

	ABILITY	MOTIVATN	ASPIRATN
ABILITY1	- -	.00	.09
ABILITY2	- -	.00	.07
ABILITY3	- -	.01	.32
MOTIVTN1	.24	- -	.02
MOTIVTN2	.96	- -	.39
MOTIVTN3	.25	- -	.52
ASPIRTN1	.11	.11	- -
ASPIRATN	.11	.11	- -

```
        EXPECTED CHANGE FOR LAMBDA-Y
```

	ABILITY	MOTIVATN	ASPIRATN
ABILITY1	- -	.00	.01
ABILITY2	- -	.00	.01
ABILITY3	- -	-.01	-.02
MOTIVTN1	.03	- -	.01
MOTIVTN2	-.07	- -	.05
MOTIVTN3	.03	- -	-.06
ASPIRTN1	.02	.05	- -
ASPIRATN	-.02	-.06	- -

```
NO NON-ZERO MODIFICATION INDICES FOR PSI

        MODIFICATION INDICES FOR THETA-EPS
```

	ABILITY1	ABILITY2	ABILITY3	MOTIVTN1	MOTIVTN2	MOTIVTN3
ABILITY1	- -					
ABILITY2	.07	- -				
ABILITY3	.01	.02	- -			

MOTIVTN1	.11	.30	2.09	- -		
MOTIVTN2	1.19	1.28	.40	.08	- -	
MOTIVTN3	1.13	1.66	.00	.18	.49	- -
ASPIRTN1	2.81	1.47	.39	1.80	2.09	7.48
ASPIRATN	4.30	.79	1.88	1.13	1.23	4.67

MODIFICATION INDICES FOR THETA-EPS

	ASPIRTN1	ASPIRATN
ASPIRTN1	- -	
ASPIRATN	- -	- -

EXPECTED CHANGE FOR THETA-EPS

	ABILITY1	ABILITY2	ABILITY3	MOTIVTN1	MOTIVTN2	MOTIVTN3
ABILITY1	- -					
ABILITY2	.01	- -				
ABILITY3	.00	.00	- -			
MOTIVTN1	-.01	-.01	.03	- -		
MOTIVTN2	.02	-.02	-.01	-.01	- -	
MOTIVTN3	-.02	.03	.00	-.01	.02	- -
ASPIRTN1	-.03	.02	.01	.03	.04	-.06
ASPIRATN	.04	-.02	-.02	-.02	-.03	.05

EXPECTED CHANGE FOR THETA-EPS

	ASPIRTN1	ASPIRATN
ASPIRTN1	- -	
ASPIRATN	- -	- -

MAXIMUM MODIFICATION INDEX IS   7.48 FOR ELEMENT (7, 6) OF THETA-EPS

In chapter 1, the topic of modification indices is discussed and it is suggested that all indices found to be larger than 5 should merit closer inspection. It is also indicated that any model changes based on modification indices should be justified on theoretical grounds and be consistent with available theories. Although there is a modification index larger than 5 in this output—the element (7,6) of the TE matrix—adding this parameter to the proposed model cannot be theoretically justified (i.e., there does not appear to be a substantive reason for the error terms of ASPIRTN1 and MOTINTN3 to correlate). In addition, because the model fit is already acceptable no change should be made to the proposed model. The model is therefore considered to be an acceptable means of data description.

TESTING MODEL RESTRICTIONS: TAU-EQUIVALENCE

The model examined in the previous section focuses on estimating the degree of interrelationship among ability, motivation, and aspiration. In examining the results, however, it appears that many of the loadings on each factor also are quite similar. Imagine now that a researcher hypothesizes that the factor loadings of the ability construct associated with the measures ABILITY1, ABILITY2, and ABILITY3 are indeed equal. Such a model is referred to in the psychometric literature as a model with tau-equivalent measures. A *tau-equivalent indicator model* amounts to the assumption that the indicators used in a study measure the same latent variable in the same units of measurement. As it turns out, all three latent variables included in this study (i.e., ability, motivation, and aspiration) can be tested for tau-equivalence by simply imposing the appropriate model restrictions on the indicators one at a time. Thus, to test for tau-equivalence of measures for each latent variable, the pertinent model restrictions are introduced and the resulting difference in chi-square values for the imposed model constraints are evaluated (for a complete discussion of the chi-square difference test, see chap. 3).

Begin by imposing equality constraints on the factor loadings of the ability construct. This restriction can be included in the EQS input file by simply adding a /CONSTRAINT section, and in the LISREL input file by adding an EQuality line as follows:

/CONSTRAINT
(V1,F1) = (V2,F1) = (V3,F1);

and

EQ LY(1, 1) LY(2, 1) LY(3, 1)

The result of introducing this restriction is an increase in the chi-square value up to $T = 25.74$ with $df = 19$ (recall that the model tested without the tau-equivalence resulted in a chi-square value of $T = 20.58$ with $df = 17$). The difference in the chi-square values between this model and the one originally proposed (displayed in Fig. 10) is therefore $\Delta T = 5.16$ for $\Delta df = 2$, and is nonsignificant. (The critical value of the chi-square distribution with 2 degrees of freedom is 5.99 at the .05 significance level, and higher for any lower level.) One can therefore conclude that the imposed factor loading identity is plausible, and hence that the ability measures are tau-equivalent. Because the restriction is found to be acceptable, it is retained in the model.

Next, test whether the motivation indicators are also tau-equivalent by imposing the identity restriction on their factor loadings. This is achieved by including the line

(V4,F2) = (V5,F2) = (V6,F2);

as a second line of the /CONSTRAINTS section in the EQS input, or by adding the line

EQ LY(4, 2) LY(5, 2) LY(6, 2)

as a second EQuality line in the LISREL input. The newly restricted model
is now associated with a chi-square value of $T = 28.56$ with $df = 21$. The
difference in chi-square values compared to the preceding model is now
$\Delta T = 2.82$, with the difference in degrees of freedom being $\Delta df = 2$, and is
hence nonsignificant. Once again, these results lead to the conclusion that
there is not enough evidence in the data to warrant the rejection of the restriction of identical factor loadings for the motivation construct, and
therefore one can consider the latter to be tau-equivalent measures of motivation. Thus, due to this nonsignificant finding, this factor loading equality is also retained in the model.

Finally, impose the restriction of tau-equivalence on the aspiration indicators. This is accomplished by adding the following line in the /CON-
STRAINTS section of the EQS input file:

(V7,F3) = (V8,F3);

and the following EQuality constraint in the LISREL input file:

EQ LY(7, 3) LY(8, 3)

The result of the last imposed restriction brings about only a slight increase in the chi-square value to $T = 28.60$ with $df = 22$. Thus, the change
in chi-square is only $\Delta T = 0.04$, for a difference in degrees of freedom of
$\Delta df = 1$, and is nonsignificant. (Recall that the critical value of the
chi-square distribution with 1 degree of freedom is 3.84 at the .05 significance level, and higher for any lower level.) Based on these results, one
can conclude that there is not enough evidence in the data to warrant rejection of the restriction of identical factor loadings for the aspiration construct, and hence one can consider these indicators to be tau-equivalent
measures of aspiration.

Thus, the above results indicate that there is not sufficient evidence in
the data to disconfirm the tau-equivalence hypothesis for any of the sets of
measures assessing the ability, motivation, and aspiration construct. One
can conclude that the indicators of ability, motivation, and aspiration assess their underlying latent variables in the same units of measurement.

Only the EQS output file for the most restricted model is presented
next, with comments inserted at appropriate places and with repetitive
material omitted (interested readers can easily verify that the results of the

LISREL and EQS programs are identical by running the appropriate LISREL input file).

```
PARAMETER ESTIMATES APPEAR IN ORDER,
NO SPECIAL PROBLEMS WERE ENCOUNTERED DURING OPTIMIZATION.

ALL EQUALITY CONSTRAINTS WERE CORRECTLY IMPOSED
```

As indicated earlier, this is an important message provided by EQS because it indicates that the program has not encountered any problems in terms of model identification and that the model constraints were correctly introduced.

RESIDUAL COVARIANCE MATRIX (S-SIGMA):

		ABILITY1	ABILITY2	ABILITY3	MOTIVTN1	MOTIVN2
		V 1	V 2	V 3	V 4	V 5
ABILITY1 V	1	-0.022				
ABILITY2 V	2	0.021	0.049			
ABILITY3 V	3	-0.029	0.021	-0.018		
MOTIVTN1 V	4	-0.028	0.002	-0.008	-0.043	
MOTIVN2 V	5	0.002	0.012	-0.018	-0.019	0.024
MOTIVN3 V	6	-0.008	0.052	0.002	-0.019	0.041
ASPIRTN1 V	7	-0.014	0.026	-0.004	-0.011	0.029
ASPIRTN2 V	8	0.016	0.006	-0.024	-0.031	0.009

		MOTIVN3	ASPIRTN1	ASPIRTN2
		V 6	V 7	V 8
MOTIVN3 V	6	0.021		
ASPIRTN1 V	7	-0.021	-0.003	
ASPIRTN2 V	8	0.029	0.000	0.003

```
            AVERAGE ABSOLUTE COVARIANCE RESIDUALS    =        0.0191
AVERAGE OFF-DIAGONAL ABSOLUTE COVARIANCE RESIDUALS   =        0.0179
```

STANDARDIZED RESIDUAL MATRIX:

		ABILITY1	ABILITY2	ABILITY3	MOTIVTN1	MOTIVN2
		V 1	V 2	V 3	V 4	V 5
ABILITY1 V	1	-0.048				
ABILITY2 V	2	0.041	0.088			
ABILITY3 V	3	-0.066	0.041	-0.041		
MOTIVTN1 V	4	-0.056	0.004	-0.015	-0.078	
MOTIVN2 V	5	0.004	0.020	-0.032	-0.031	0.036
MOTIVN3 V	6	-0.015	0.090	0.004	-0.033	0.065
ASPIRTN1 V	7	-0.028	0.046	-0.008	-0.019	0.047
ASPIRTN2 V	8	0.030	0.010	-0.046	-0.053	0.015

		MOTIVN3	ASPIRTN1	ASPIRTN2
		V 6	V 7	V 8
MOTIVN3 V	6	0.035		
ASPIRTN1 V	7	-0.035	-0.005	
ASPIRTN2 V	8	0.048	0.000	0.005

```
              AVERAGE ABSOLUTE STANDARDIZED RESIDUALS  =        0.0344
    AVERAGE OFF-DIAGONAL ABSOLUTE STANDARDIZED RESIDUALS  =        0.0321
```

LARGEST STANDARDIZED RESIDUALS:

```
  V   6,V   2     V   2,V   2     V   4,V   4     V   3,V   1     V   6,V   5
     0.090           0.088          -0.078          -0.066           0.065

  V   4,V   1     V   8,V   4     V   1,V   1     V   8,V   6     V   7,V   5
    -0.056          -0.053          -0.048           0.048           0.047

  V   8,V   3     V   7,V   2     V   3,V   3     V   2,V   1     V   3,V   2
    -0.046           0.046          -0.041           0.041           0.041

  V   5,V   5     V   6,V   6     V   7,V   6     V   6,V   4     V   5,V   3
     0.036           0.035          -0.035          -0.033          -0.032
```

DISTRIBUTION OF STANDARDIZED RESIDUALS

```
     - - - - - - - - - - - - - - - - - - - - - - - - - - -
     !                                      !
  20-                                       -
     !                     *                !
     !                     *                !
     !                 *   *                !
     !                 *   *                !        RANGE        FREQ  PERCENT
  15-                  *   *                -
     !                 *   *                !    1  -0.5  -  --       0   0.00%
     !                 *   *                !    2  -0.4  -  -0.5     0   0.00%
     !                 *   *                !    3  -0.3  -  -0.4     0   0.00%
     !                 *   *                !    4  -0.2  -  -0.3     0   0.00%
  10-                  *   *                -    5  -0.1  -  -0.2     0   0.00%
     !                 *   *                !    6   0.0  -  -0.1    17  47.22%
     !                 *   *                !    7   0.1  -   0.0    19  52.78%
     !                 *   *                !    8   0.2  -   0.1     0   0.00%
     !                 *   *                !    9   0.3  -   0.2     0   0.00%
   5-                  *   *                -    A   0.4  -   0.3     0   0.00%
     !                 *   *                !    B   0.5  -   0.4     0   0.00%
     !                 *   *                !    C   ++   -   0.5     0   0.00%
     !                 *   *                !      - - - - - - - - - - - - - - -
     !                 *   *                !          TOTAL        36 100.00%
     - - - - - - - - - - - - - - - - - - - - - - -
       1  2  3  4  5  6  7  8  9  A  B  C    EACH "*" REPRESENTS  1 RESIDUALS
```

Once again, even with the imposed restrictions there are no outstanding residuals, and hence no signs that the model does not fit well.

```
GOODNESS OF FIT SUMMARY
INDEPENDENCE MODEL CHI-SQUARE =         801.059 ON    28 DEGREES OF FREEDOM
INDEPENDENCE AIC =     745.05881   INDEPENDENCE CAIC =    618.45790
         MODEL AIC =     -15.40191          MODEL CAIC =   -114.87405
```

```
CHI-SQUARE =        28.598 BASED ON     22 DEGREES OF FREEDOM
PROBABILITY VALUE FOR THE CHI-SQUARE STATISTIC IS       0.15670
THE NORMAL THEORY RLS CHI-SQUARE FOR THIS ML SOLUTION IS              26.952.

BENTLER-BONETT NORMED    FIT INDEX=    0.964
BENTLER-BONETT NONNORMED FIT INDEX=    0.989
COMPARATIVE FIT INDEX            =    0.991
```

All the goodness-of-fit indices suggest that the model fits the data well and hence that interpretations of its parameter estimates can be trusted.

```
                         ITERATIVE SUMMARY

                      PARAMETER
ITERATION             ABS CHANGE          ALPHA           FUNCTION
     1                0.308560          1.00000          0.56251
     2                0.078387          1.00000          0.15343
     3                0.017454          1.00000          0.11493
     4                0.000874          1.00000          0.11485
```

This is, again, a clear sign of a quick and clean convergence to the final solution.

```
MEASUREMENT EQUATIONS WITH STANDARD ERRORS AND TEST STATISTICS

ABILITY1=V1    =       .547*F1     + 1.000 E1
                       .030
                     18.419

ABILITY2=V2    =       .547*F1     + 1.000 E2
                       .030
                     18.419

ABILITY3=V3    =       .547*F1     + 1.000 E3
                       .030
                     18.419

MOTIVTN1=V4    =       .565*F2     + 1.000 E4
                       .033
                     16.962

MOTIVN2 =V5    =       .565*F2     + 1.000 E5
                       .033
                     16.962

MOTIVN3 =V6    =       .565*F2     + 1.000 E6
                       .033
                     16.962

ASPIRTN1=V7    =       .624*F3     + 1.000 E7
                       .036
                     17.208

ASPIRTN2=V8    =       .624*F3     + 1.000 E8
                       .036
                     17.208
```

In this model, all sets of indicators are constrained for tau-equivalence (i.e., there is factor-loading identity within each set of indicators), thereby leading to identical standard errors and t values within sets.

```
VARIANCES OF INDEPENDENT VARIABLES
----------------------------------
              V                                       F
             ---                                     ---
                             I F1 -ABILITY       1.000 I
                             I                         I
                             I                         I
                             I                         I
                             I F2 -MOTIVATN      1.000 I
                             I                         I
                             I                         I
                             I                         I
                             I F3 -ASPIRATN      1.000 I
                             I                         I
                             I                         I
                             I                         I

VARIANCES OF INDEPENDENT VARIABLES
----------------------------------

              E                             D
             ---                           ---
E1 -ABILITY1               .172*I                         I
                          .022 I                          I
                         7.974 I                          I
                               I                          I
E2 -ABILITY2               .211*I                         I
                          .025 I                          I
                         8.577 I                          I
                               I                          I
E3 -ABILITY3               .169*I                         I
                          .021 I                          I
                         7.914 I                          I
                               I                          I
E4 -MOTIVTN1               .274*I                         I
                          .031 I                          I
                         8.735 I                          I
                               I                          I
E5 -MOTIVN2                .317*I                         I
                          .035 I                          I
                         9.087 I                          I
                               I                          I
E6 -MOTIVN3                .270*I                         I
                          .031 I                          I
                         8.695 I                          I
                               I                          I
E7 -ASPIRTN1               .193*I                         I
                          .030 I                          I
                         6.458 I                          I
                               I                          I
```

```
E8 -ASPIRTN2                        .227*I                                    I
                                    .032 I                                    I
                                  7.146 I                                     I
                                        I                                     I


COVARIANCES AMONG INDEPENDENT VARIABLES
------------------------------------.

            V                                           F
            ---                                         ---
                                 I  F2 -MOTIVATN              .640*I
                                 I  F1 -ABILITY              .055 I
                                 I                         11.687 I
                                 I                                I
                                 I  F3 -ASPIRATN              .275*I
                                 I  F1 -ABILITY              .074 I
                                 I                          3.738 I
                                 I                                I
                                 I  F3 -ASPIRATN              .683*I
                                 I  F2 -MOTIVATN             .054 I
                                 I                         12.590 I
```

These are the factor correlation estimates along with their standard errors and t values. The estimates suggest the same conclusions that were obtained for the unconstrained model concerning the interrelationships between the latent variables in the population. (Recall, however, that, in general, conclusions based on more restrictive tenable models are more precise, as elaborated on in earlier sections of the book.) Thus, the presence of tau-equivalence in the proposed model does not lead to different substantive conclusions.

CHAPTER FIVE

Structural Regression Models

WHAT IS A STRUCTURAL REGRESSION MODEL?

In chapter 4 the confirmatory factor analysis model was presented as a method for examining the patterns of interrelationships among several constructs. No specific directional relationships were assumed among the constructs, only that they were correlated with one another. And yet, in many scientific fields of study, models that postulate specific explanatory relationships (regressions) among constructs are often proposed. To set these models apart from other models discussed in this book, we will refer to them in this chapter as structural regression models.

Although structural regression models resemble confirmatory factor analysis models, they possess the noteworthy characteristic that some of their latent variables (i.e., elements of the structure of the phenomenon under investigation) are regressed on others. Thus, once the constructs have been assessed, structural regression models can be used to test the plausibility of hypothetical assertions about their explanatory relationships. In terms of their path diagrams, structural regression models will always have at least one path (one-way arrow) leaving a putative explanatory latent variable and ending at another construct. Thus, in a way, structural regression models can be viewed as extensions of path analysis models (discussed in chap. 3) except that, instead of being conceived in terms of only observed variables, the models also include latent variables.

EXAMPLE STRUCTURAL REGRESSION MODEL

To demonstrate a structural regression model, consider the following example concerning mental ability. General mental ability is one of the most extensively studied constructs in the behavioral sciences. According to one popular theory (e.g., Horn, 1982), human intellectual capabilities can be roughly classified into two main clusters, fluid and crystallized intelligence. *Fluid intelligence* is the component of general intelligence that reflects people's ability to quickly process a potentially large amount of information in order to solve content-free tasks based on contexts that they are not familiar with from their education or prior socialization process. Metaphorically, fluid intelligence can be thought of as resembling a human brain's hardware, or the mechanics of a brain. Fluid intelligence does not include people's abilities to retrieve knowledge obtained earlier in life through systems of culture or education, but instead refers to their ability to create knowledge when solving unfamiliar problems. *Crystallized intelligence*, on the other hand, is people's ability to retrieve knowledge probably obtained earlier in life through systems of culture and education. In pragmatic terms, tests of fluid intelligence frequently contain series of context-free symbols that are arranged following a special rule that must be discovered by subjects and subsequently used by them in order to arrive at a correct solution. Alternatively, measures of crystallized intelligence typically contain items that assess subjects' levels of knowledge in certain areas.

The example structural regression model considered here focuses on two particular fluid-intelligence components, induction and figural relations. Induction relates to people's ability to reason using analogies and rules of induction to more general contexts. Figural relations pertains to people's ability to see patterns of relationships between parts of figures, mentally rotate them, and also use forms of inductive reasoning with figural elements. A total of nine measures were collected from a sample of $N = 220$ high school students. The following observed variables were used in the study:

1. Induction score 1 obtained in junior year (IND1).
2. Induction score 2 obtained in junior year (IND2).
3. Induction score 3 obtained in junior year (IND3).
4. Figural relations score 1 obtained in junior year (FR11).
5. Figural relations score 2 obtained in junior year (FR12).
6. Figural relations score 3 obtained in junior year (FR13).
7. Figural relations score 1 obtained in senior year (FR21).
8. Figural relations score 2 obtained in senior year (FR22).
9. Figural relations score 3 obtained in senior year (FR23).

The example structural regression model is presented in Fig. 11 and the observed covariance matrix is presented in Table 2. The model is initially formulated in LISREL notation using Y_1 to Y_9 for the observed variables, ε_1

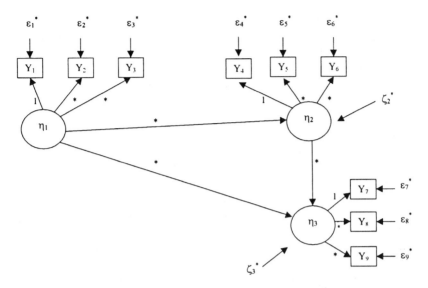

FIG. 11. Example structural regression model using LISREL notation.

to ε_9 for the error terms associated with the observed variables, and η_1 to η_3 for the latent variables. The proposed model assumes that the Induction construct (measured during the junior year of high school) plays an explanatory role for the Figural relations construct (measured during both the junior and senior years of high school). In addition, the model posits that a student's junior-year Figural relations affects his or her senior-year Figural relations.

To determine the parameters of the model presented in Fig. 11 (designated by asterisks), one must again follow the six rules outlined in chapter

TABLE 2
Covariance Matrix for the Structural Regression Model Example

Variable	IND1	IND2	IND3	FR11	FR12	FR13	FR21	FR22	FR23
IND1	56.21								
IND2	31.55	75.55							
IND3	23.27	28.30	44.45						
FR11	24.48	32.24	22.56	84.64					
FR12	22.51	29.54	20.61	57.61	78.93				
FR13	22.65	27.56	15.33	53.57	49.27	73.76			
FR21	33.24	46.49	31.44	67.81	54.76	54.58	141.77		
FR22	32.56	40.37	25.58	55.82	52.33	47.74	98.62	117.33	
FR23	30.32	40.44	27.69	54.78	53.44	59.52	96.95	84.87	106.35

Notes. IND*i* = *i*th INDuction indicator at first assessment; FR1*i* = *i*th Figural Relations indicator at first assessment, and FR2*i* = *i*th Figural Relations indicator at second assessment (i = 1, 2, 3).

1. According to Rule 1, all nine error term variances are model parameters and, according to Rule 3, the nine factor loadings are also model parameters. In addition, the variances of the structural regression disturbance terms ζ_2 and ζ_3 are also model parameters. These disturbances pertain to the latent variables η_2 and η_3, respectively (i.e., the junior- and senior-year Figural relations). Thereby, ζ_2 represents the part of junior-year Figural relations that is not accounted for in terms of its postulated (linear) explanatory relationship to Induction. Similarly, ζ_3 stands for the part of senior-year Figural relations that is not explained in terms of its assumed (linear) relationships to Induction and junior-year Figural relations. Note that Rule 2 is not applicable in this model because it does not have any proposed latent covariances. Indeed, those between Induction and both Figural relation constructs are explained in terms of their structural regression coefficients, which, following Rule 4, are found to be model parameters. Following the LISREL notation discussed in chapter 2, they can be denoted as β_{21}, β_{31}, and β_{32}. These parameters therefore relate Induction (η_1) to junior-year Figural relations (η_2) and senior-year Figural relations (η_3), and junior-year Figural relations (η_2) to senior-year Figural relations (η_3). Finally, Rule 6 requires that the scale of each latent variable be fixed. Because this study is focused on determining the explanatory role of the Induction and junior-year Figural relations constructs, it is easier to achieve the latent scale fixing by simply setting the loading of the first indicator on each latent variable to 1. Using this approach ensures that the Figural relations construct is assessed in the same metric on both occasions. Thus, the model in Fig. 11 has, altogether, 21 parameters symbolized by asterisks.

LISREL AND EQS INPUT FILES

EQS Input File

The EQS input file shown next includes two new definitions. The first definition deals with the specification of the equations that relate the latent variables in the model to one another. Because the model presented in Fig. 11 includes structural regressions, the following two equations must be included in the input file: (a) F2 = *F1 + D2, and (b) F3 = *F1 + *F2 + D3. The second definition concerns the two structural-disturbance terms included in the model (i.e., the D2 and D3 in the equations; these terms correspond to the residual variables ζ_2 and ζ_3 in the LISREL input file presented later). Because D2 and D3 are independent variables, their variances are model parameters as indicated in the /VARIANCE section. Thus, the following EQS input file is created:

```
/TITLE
EXAMPLE STRUCTURAL REGRESSION MODEL;
/SPECIFICATIONS
CASES=220; VARIABLES=9;
/LABELS
V1=IND1; V2=IND2; V3=IND3; V4=FR11; V5=FR12; V6=FR13;
V7=FR21; V8= FR22; V9=FR23;
F1=INDUCTN; F2=FIGREL1; F3=FIGREL2;
/EQU
V1= F1+E1;
V2=*F1+E2;
V3=*F1+E3;
V4= F2+E4;
V5=*F2+E5;
V6=*F2+E6;
V7= F3+E7;
V8=*F3+E8;
V9=*F3+E9;
F2=*F1+D2;
F3=*F1+*F2+D3;
/VAR
F1=*; D2 TO D3=*; E1 TO E9=*;
/MATRIX
56.21
31.55  75.55
23.27  28.30  44.45
24.48  32.24  22.56  84.64
22.51  29.54  20.61  57.61  78.93
22.65  27.56  15.33  53.57  49.27  73.76
33.24  46.49  31.44  67.81  54.76  54.58  141.77
32.56  40.37  25.58  55.82  52.33  47.74  98.62  117.33
30.32  40.44  27.69  54.78  53.44  59.52  96.95  84.87  106.35
/END;
```

LISREL Input File

The LISREL input file is constructed following the principles outlined in chapter 2. The file begins with a title command line followed by information about the number of analyzed variables, sample size and covariance matrix, and observed variable labels. Finally, in the model command line, the matrices PS, TE, LY, and BE are declared. Thus, the following LISREL input file is created:

```
EXAMPLE STRUCTURAL REGRESSION MODEL
DA NI=9 NO=220
CM
56.21
31.55   75.55
23.27   28.30   44.45
24.48   32.24   22.56   84.64
22.51   29.54   20.61   57.61   78.93
22.65   27.56   15.33   53.57   49.27   73.76
33.24   46.49   31.44   67.81   54.76   54.58   141.77
32.56   40.37   25.58   55.82   52.33   47.74   98.62    117.33
30.32   40.44   27.69   54.78   53.44   59.52   96.95    84.87   106.35
LA
IND1 IND2 IND3 FR11 FR12 FR13 FR21 FR22 FR23
MO NY=9 NE=3 PS=SY,FI TE=DI,FR LY=FU,FI BE=FU,FI
LE
INDUCTN FIGREL1 FIGREL2
FR LY(2, 1) LY(3, 1)
FR LY(5, 2) LY(6, 2)
FR LY(8, 3) LY(9, 3)
VA 1 LY(1, 1) LY(4, 2) LY(7, 3)
FR BE(2, 1) BE(3, 1) BE(3, 2)
FR PS(1, 1) PS(2, 2) PS(3, 3)
OU
```

As can be seen, the covariance matrix of the Induction construct and the two structural disturbances associated with Figural relations are initially declared to be fixed (in order to fix their covariances), but are freed in the two lines right before the OUtput command. The structural regression matrix is also declared to be fixed and only those elements of interest according to the proposed model (i.e., β_{21}, β_{31}, and β_{32}) are defined as free. This strategy of fixing everything first to 0 and then freeing those elements of interest is a more economical way of setting up the input file. Similarly, the factor loading matrix LY is declared to be fixed and in the next three lines only those of its elements that pertain to the six factor loadings to be estimated are defined as free. Subsequently, in the following line, the latent variable scales are fixed by setting the first indicator loading to a value of 1. Finally, by declaring the error covariance matrix (TE) to be diagonal and free in the model definition line, all its nine elements along the main diagonal are effectively defined as parameters (i.e., the nine error-term variances).

MODELING RESULTS

EQS Program Results

The output produced by the EQS input file created in the previous section is presented next (interested readers can easily verify that the results of the LISREL and EQS programs are identical by running the appropriate LISREL input file). As before, comments are inserted at appropriate places to clarify portions of the output. In addition, the echoed input file, the analyzed covariance matrix, and the recurring title line are omitted.

```
PARAMETER ESTIMATES APPEAR IN ORDER,
NO SPECIAL PROBLEMS WERE ENCOUNTERED DURING OPTIMIZATION.
```

As indicated in previous chapters, this message is a reassurance that the program has not encountered problems stemming from lack of identification or other numerical difficulties and that the model is technically sound.

```
RESIDUAL COVARIANCE MATRIX (S-SIGMA):
```

			IND1 V 1	IND2 V 2	IND3 V 3	FR11 V 4	FR12 V 5
IND1	V	1	0.000				
IND2	V	2	-0.552	0.000			
IND3	V	3	0.790	-0.210	0.000		
FR11	V	4	-0.227	0.905	0.617	0.000	
FR12	V	5	-0.165	0.783	0.472	0.882	0.000
FR13	V	6	1.007	0.111	-3.891	-0.576	-0.423
FR21	V	7	-2.149	1.609	0.011	2.579	-5.106
FR22	V	8	1.487	0.962	-2.016	-1.457	-0.236
FR23	V	9	-0.889	0.859	-0.027	-2.748	0.644

			FR13 V 6	FR21 V 7	FR22 V 8	FR23 V 9
FR13	V	6	0.000			
FR21	V	7	-2.562	0.000		
FR22	V	8	-2.434	1.631	0.000	
FR23	V	9	9.127	-0.464	-0.665	0.000

```
                    AVERAGE ABSOLUTE COVARIANCE RESIDUALS         =        1.1394
        AVERAGE OFF-DIAGONAL ABSOLUTE COVARIANCE RESIDUALS        =        1.4242
STANDARDIZED RESIDUAL MATRIX:
```

			IND1 V 1	IND2 V 2	IND3 V 3	FR11 V 4	FR12 V 5
IND1	V	1	0.000				
IND2	V	2	-0.008	0.000			
IND3	V	3	0.016	-0.004	0.000		

FR11	V	4	-0.003	0.011	0.010	0.000	
FR12	V	5	-0.002	0.010	0.008	0.011	0.000
FR13	V	6	0.016	0.001	-0.068	-0.007	-0.006
FR21	V	7	-0.024	0.016	0.000	0.024	-0.048
FR22	V	8	0.018	0.010	-0.028	-0.015	-0.002
FR23	V	9	-0.012	0.010	0.000	-0.029	0.007

			FR13	FR21	FR22	FR23
			V 6	V 7	V 8	V 9
FR13	V	6	0.000			
FR21	V	7	-0.025	0.000		
FR22	V	8	-0.026	0.013	0.000	
FR23	V	9	0.103	-0.004	-0.006	0.000

```
             AVERAGE ABSOLUTE STANDARDIZED RESIDUALS        =        0.0134
     AVERAGE OFF-DIAGONAL ABSOLUTE STNADARDIZED RESIDUALS   =        0.0167
```

LARGEST STANDARDIZED RESIDUALS:

V 9,V 6	V 6,V 3	V 7,V 5	V 9,V 4	V 8,V 3
0.103	-0.068	-0.048	-0.029	-0.028
V 8,V 6	V 7,V 6	V 7,V 1	V 7,V 4	V 8,V 1
-0.026	-0.025	-0.024	0.024	0.018
V 3,V 1	V 6,V 1	V 7,V 2	V 8,V 4	V 8,V 7
0.016	0.016	0.016	-0.015	0.013
V 9,V 1	V 4,V 2	V 5,V 4	V 8,V 2	V 5,V 2
-0.012	0.011	0.011	0.010	0.010

DISTRIBUTION OF STANDARDIZED RESIDUALS

```
     - - - - - - - - - - - - - - - - - - - - - - - - - -
     !                                      !
40 - !                                      -
     !                                      !
     !     .                                !
     !                                      !
     !                                      !      RANGE       FREQ   PERCENT
30 - !                                      -
     !                                      !  1  -0.5  -  --      0    0.00%
     !                                      !  2  -0.4  -  -0.5    0    0.00%
     !           *                          !  3  -0.3  -  -0.4    0    0.00%
     !           *  *                       !  4  -0.2  -  -0.3    0    0.00%
20 - !           *  *                       -  5  -0.1  -  -0.2    0    0.00%
     !           *  *                       !  6   0.0  -  -0.1   23   51.11%
     !           *  *                       !  7   0.1  -   0.0   21   46.67%
     !           *  *                       !  8   0.2  -   0.1    1    2.22%
     !           *  *                       !  9   0.3  -   0.2    0    0.00%
10 - !           *  *                       -  A   0.4  -   0.3    0    0.00%
     !           *  *                       !  B   0.5  -   0.4    0    0.00%
     !           *  *                       !  C   ++   -   0.5    0    0.00%
     !           *  *                       !  - - - - - - - - - - - - - - - - - - -
     !           *  *  *                     !       TOTAL        45  100.00%
     - - - - - - - - - - - - - - - - - - - - - - - - - -
        1  2  3  4  5  6  7  8  9  A  B  C   EACH "*" REPRESENTS  2 RESIDUALS
```

Based on this section of the output, it appears that there is a residual indicating a potentially important misfit of the model. It is first stated in the section LARGEST STANDARDIZED RESIDUALS and also appears to the right of the remaining residuals in the graphical distribution of standardized residuals. This large residual is associated with the last indicator of Figural relations (i.e., the third measure of Figural relations taken during the junior and senior high school years). In fact, this residual is nearly two times larger than the next one, whereas the residuals following gradually trail off to 0. Now scrutinize the model fit indices to see if this apparent deficiency may have also contributed to a lack of overall model fit.

```
GOODNESS OF FIT SUMMARY

INDEPENDENCE MODEL CHI-SQUARE =        1177.363 ON    36 DEGREES OF FREEDOM

INDEPENDENCE AIC =   1105.36282    INDEPENDENCE CAIC =   947.19223
          MODEL AIC =      4.09682          MODEL CAIC =  -101.35024

CHI-SQUARE =        52.097 BASED ON    24 DEGREES OF FREEDOM
PROBABILITY VALUE FOR THE CHI-SQUARE STATISTIC IS LESS THAN 0.001
THE NORMAL THEORY RLS CHI-SQUARE FOR THIS ML SOLUTION IS          48.276.

BENTLER-BONETT NORMED    FIT INDEX=      0.956
BENTLER-BONETT NONNORMED FIT INDEX=      0.963
COMPARATIVE FIT INDEX            =       0.975
```

The goodness-of-fit indices are not satisfactory and suggest that this is not a well-fitting model. For example, the chi-square value and its p value are quite unsatisfactory. Based on this, one can conclude that there is evidence in the data that the fitted model is not a reasonably good representation, neither overall nor locally at the level of variances and covariances. One therefore should not fully trust the remaining results. The rest of the output is presented next only for reasons of completeness.

```
                    ITERATIVE SUMMARY

                  PARAMETER
ITERATION         ABS CHANGE         ALPHA          FUNCTION
    1             23.169935        1.00000          16.31806
    2             14.960183        1.00000          10.47132
    3              3.720057        1.00000           7.35567
    4              4.876157        1.00000          11.22322
    5             22.961660        0.50000           6.80336
    6             58.645729        0.12500           7.55145
    7             47.166233        0.25000           7.11284
    8              9.357183        1.00000           6.04928
    9             10.148293        1.00000           4.57749
   10              3.072371        1.00000           3.25892
```

11	1.838869	1.00000	1.95995
12	3.086363	1.00000	0.78450
13	3.693515	1.00000	0.28511
14	1.354367	1.00000	0.23848
15	0.154908	1.00000	0.23789
16	0.019709	1.00000	0.23789
17	0.004465	1.00000	0.23789
18	0.001110	1.00000	0.23788
19	0.000270	1.00000	0.23789

After an initial search in the wrong direction, the program finds a path leading to the final reported solution.

```
MEASUREMENT EQUATIONS WITH STANDARD ERRORS AND TEST STATISTICS

    IND1  =V1  =    1.000 F1    + 1.000 E1


    IND2  =V2  =    1.268*F1    + 1.000 E2
                     .157
                    8.084

    IND3  =V3  =     .888*F1    + 1.000 E3
                     .115
                    7.701

    FR11  =V4  =    1.000 F2    + 1.000 E4


    FR12  =V5  =     .918*F2    + 1.000 E5
                     .067
                   13.762

    FR13  =V6  =     .876*F2    + 1.000 E6
                     .065
                   13.540

    FR21  =V7  =    1.000 F3    + 1.000 E7


    FR22  =V8  =     .878*F3    + 1.000 E8
                     .052
                   16.790

    FR23  =V9  =     .882*F3    + 1.000 E9
                     .048
                   18.393


CONSTRUCT EQUATIONS WITH STANDARD ERRORS AND TEST STATISTICS

FIGREL1 =F2  =     .976*F1    + 1.000 D2
                    .147
                   6.642
FIGREL2 =F3  =     .814*F2    + .603*F1    + 1.000 D3
                    .110         .177
                   7.397        3.405
```

```
VARIANCES OF INDEPENDENT VARIABLES
----------------------------------

                   V                              F
                   ---                            ---
                                 I F1 -INDUCTN        25.312*I
                                 I                     5.144 I
                                 I                     4.921 I
                                 I                           I
                   E                              D
                   ---                            ---
E1 - IND1               30.898*I D2 -FIGREL1         37.695*I
                         3.878 I                      6.101 I
                         7.968 I                      6.179 I
                               I                            I
E2 - IND2               34.837*I D3 -FIGREL2         35.998*I
                         5.058 I                      5.923 I
                         6.887 I                      6.078 I
                               I                            I
E3 - IND3               24.486*I                            I
                         3.068 I                            I
                         7.980 I                            I
                               I                            I
E4 - FR11               22.828*I                            I
                         3.423 I                            I
                         6.670 I                            I
                               I                            I
E5 - FR12               26.868*I                            I
                         3.468 I                            I
                         7.747 I                            I
                               I                            I
E6 - FR13               26.329*I                            I
                         3.313 I                            I
                         7.947 I                            I
                               I                            I
E7 - FR21               31.311*I                            I
                         4.400 I                            I
                         7.116 I                            I
                               I                            I
E8 - FR22               32.168*I                            I
                         4.025 I                            I
                         7.993 I                            I
                               I                            I
E9 - FR23               20.441*I                            I
                         3.145 I                            I
                         6.499 I                            I
                               I                            I


STANDARDIZED SOLUTION:


  IND1  =V1  =    .671 F1    +  .741 E1
  IND2  =V2  =    .734*F1    +  .679 E2
  IND3  =V3  =    .670*F1    +  .742 E3
  FR11  =V4  =    .855 F2    +  .519 E4
```

```
    FR12 =V5  =     .812*F2   +   .583 E5
    FR13 =V6  =     .802*F2   +   .597 E6
    FR21 =V7  =     .883 F3   +   .470 E7
    FR22 =V8  =     .852*F3   +   .524 E8
    FR23 =V9  =     .899*F3   +   .438 E9
FIGREL1 =F2  =     .625*F1   +   .781 D2
FIGREL2 =F3  =     .609*F2   +   .289*F1   +   .571 D3
```

This is the final model solution presented in nearly the same form as the model equations provided to EQS in the input file. However, because the model was found not to be a good means of data description, one should not attempt any interpretation of the parameter estimates. Instead, one should be more concerned with finding a way to improve the model fit. Because a large positive standardized residual corresponding to a particular part of the model has been found, perhaps the model can be improved by making some modifications. In particular, the large positive standardized residual suggests that the model underpredicts the relationship between the repeated assessments with the third Figural relation measure. With multiple measurements involving the same set of indicators, it is possible that the error terms associated with the measures contain some specific variance that is not explained by the latent variables. This seems to be the most likely scenario in the present case with regard to the last indicator within the fluid measures used. Note that in the previous chapter the issue of trying to make only modifications to models that can be theoretically justified was discussed. Here there does appear to be a good substantive reason for the error terms of the Figural relations measures FR13 and FR23 to correlate. Therefore this particular covariance between the error terms in the next model version is freed and added to the EQS input file as

/COVARIANCE
E9,E6=*;

The pertinent parts of the resulting output are presented next.

```
PARAMETER ESTIMATES APPEAR IN ORDER,
NO SPECIAL PROBLEMS WERE ENCOUNTERED DURING OPTIMIZATION.

RESIDUAL COVARIANCE MATRIX (S-SIGMA):
```

			IND1	IND2	IND3	FR11	FR12
			V 1	V 2	V 3	V 4	V 5
IND1	V	1	0.000				
IND2	V	2	-0.429	0.000			
IND3	V	3	0.764	-0.300	0.000		
FR11	V	4	-0.649	0.308	0.087	0.000	

FR12	V	5	0.199	1.188	0.656	-0.078	0.000
FR13	V	6	1.710	0.950	-3.397	-0.573	1.197
FR21	V	7	-2.459	1.126	-0.486	2.300	-3.406
FR22	V	8	1.332	0.686	-2.348	-1.487	1.448
FR23	V	9	-0.516	1.255	0.113	-1.807	3.197

			FR13	FR21	FR22	FR23
			V 6	V 7	V 8	V 9
FR13	V	6	0.101			
FR21	V	7	-0.011	0.000		
FR22	V	8	-0.015	0.322	0.000	
FR23	V	9	0.101	-0.113	-0.038	0.007

AVERAGE ABSOLUTE COVARIANCE RESIDUALS   =   0.8257
AVERAGE OFF-DIAGONAL ABSOLUTE COVARIANCE RESIDUALS   =   1.0292

STANDARDIZED RESIDUAL MATRIX:

			IND1	IND2	IND3	FR11	FR12
			V 1	V 2	V 3	V 4	V 5
IND1	V	1	0.000				
IND2	V	2	-0.007	0.000			
IND3	V	3	0.015	-0.005	0.000		
FR11	V	4	-0.009	0.004	0.001	0.000	
FR12	V	5	0.003	0.015	0.011	-0.001	0.000
FR13	V	6	0.027	0.013	-0.059	-0.007	0.016
FR21	V	7	-0.028	0.011	-0.006	0.021	-0.032
FR22	V	8	0.016	0.007	-0.033	-0.015	0.015
FR23	V	9	-0.007	0.014	0.002	-0.019	0.035

			FR13	FR21	FR22	FR23
			V 6	V 7	V 8	V 9
FR13	V	6	0.001			
FR21	V	7	0.000	0.000		
FR22	V	8	0.000	0.002	0.000	
FR23	V	9	0.001	-0.001	0.000	0.000

AVERAGE ABSOLUTE STANDARDIZED RESIDUALS   =   0.0102
AVERAGE OFF-DIAGONAL ABSOLUTE STANDARDIZED RESIDUALS   =   0.0127

LARGEST STANDARDIZED RESIDUALS:

V 6,V 3	V 9,V 5	V 8,V 3	V 7,V 5	V 7,V 1
-0.059	0.035	-0.033	-0.032	-0.028

V 6,V 1	V 7,V 4	V 9,V 4	V 8,V 1	V 6,V 5
0.027	0.021	-0.019	0.016	0.016

V 5,V 2	V 3,V 1	V 8,V 5	V 8,V 4	V 9,V 2
0.015	0.015	0.015	-0.015	0.014

V 6,V 2	V 5,V 3	V 7,V 2	V 4,V 1	V 8,V 2
0.013	0.011	0.011	-0.009	0.007

DISTRIBUTION OF STANDARDIZED RESIDUALS

```
     - - - - - - - - - - - - - - - - - - - - - - - - - -
     !                                  !
 40 -                                   -
     !                                  !
     !                                  !
     !                                  !
     !                                  !         RANGE        FREQ  PERCENT
 30 -                                   -
     !                                  !     1 . -0.5  -   --       0    0.00%
     !                                  !     2  -0.4  - -0.5        0    0.00%
     !                   *              !     3  -0.3  - -0.4        0    0.00%
     !              *    *              !     4  -0.2  - -0.3        0    0.00%
 20 -              *    *               -     5  -0.1  - -0.2        0    0.00%
     !              *    *              !     6   0.0  - -0.1       21   46.67%
     !              *    *              !     7   0.1  -  0.0       24   53.33%
     !              *    *              !     8   0.2  -  0.1        0    0.00%
     !              *    *              !     9   0.3  -  0.2        0    0.00%
 10 -              *    *               -     A   0.4  -  0.3        0    0.00%
     !              *    *              !     B   0.5  -  0.4        0    0.00%
     !              *    *              !     C   ++   -  0.5        0    0.00%
     !              *    *              !     - - - - - - - - - - - - - - - - - - - - -
     !              *    *              !         TOTAL            45  100.00%
     - - - - - - - - - - - - - - - - - - - - - - - - - -
       1  2  3  4  5  6  7  8  9  A  B  C    EACH "*" REPRESENTS  2 RESIDUALS
```

None of the residuals presented in this section of the output are a cause for concern. As indicated previously, this is typically the case for well-fitting models with regard to all variable variances and covariances. The residuals listed in the LARGEST STANDARDIZED RESIDUALS section are much more uniform and smaller in magnitude than in the first model tested, and the distribution of standardized residuals is symmetric. The overall-model-fit indices are presented next.

```
GOODNESS OF FIT SUMMARY

INDEPENDENCE MODEL CHI-SQUARE =        1177.363 ON    36 DEGREES OF FREEDOM

INDEPENDENCE AIC =  1105.36282    INDEPENDENCE CAIC =  947.19223
        MODEL AIC =   -25.44765          MODEL CAIC = -126.50108

CHI-SQUARE =       20.552 BASED ON    23 DEGREES OF FREEDOM
PROBABILITY VALUE FOR THE CHI-SQUARE STATISTIC IS      0.60840
THE NORMAL THEORY RLS CHI-SQUARE FOR THIS ML SOLUTION IS         20.011.

BENTLER-BONETT NORMED    FIT INDEX=       0.983
BENTLER-BONETT NONNORMED FIT INDEX=       1.003
COMPARATIVE FIT INDEX            =        1.000
```

The goodness-of-fit indices are now quite satisfactory. Note in particular that the chi-square value is comparable to its degrees of freedom and the associated p value is well in excess of any reasonable significance level. In addition, the normed and nonnormed fit indices, and comparative fit indices are all close to 1—the NNFI is actually slightly above 1, which sometimes happens with well-fitting models. This model is therefore a reasonably good means of data description.

ITERATIVE SUMMARY

ITERATION	PARAMETER ABS CHANGE	ALPHA	FUNCTION
1	22.842024	1.00000	16.63225
2	17.548189	1.00000	10.42740
3	3.001143	1.00000	7.31791
4	4.892436	1.00000	11.32526
5	27.355335	0.50000	8.74595
6	39.696941	1.00000	8.46845
7	16.204002	1.00000	7.42914
8	41.607319	1.00000	5.72652
9	11.167980	1.00000	3.35232
10	0.927136	1.00000	2.01023
11	2.364902	1.00000	0.87052
12	4.226802	1.00000	0.18126
13	1.303813	1.00000	0.09465
14	0.105986	1.00000	0.09385
15	0.011150	1.00000	0.09385
16	0.001340	1.00000	0.09385
17	0.000186	1.00000	0.09385

MEASUREMENT EQUATIONS WITH STANDARD ERRORS AND TEST STATISTICS

```
IND1  =V1  =    1.000 F1    + 1.000 E1


IND2  =V2  =    1.271*F1    + 1.000 E2
                 .157
                8.073

IND3  =V3  =     .894*F1    + 1.000 E3
                 .116
                7.713

FR11  =V4  =    1.000 F2    + 1.000 E4


FR12  =V5  =     .888*F2    + 1.000 E5
                 .064
               13.885

FR13  =V6  =     .833*F2    + 1.000 E6
                 .062
               13.465
```

```
FR21 =V7   =    1.000 F3    + 1.000 E7


FR22 =V8   =     .875*F3    + 1.000 E8
                 .051
               17.202

FR23 =V9   =     .864*F3    + 1.000 E9
                 .047
               18.391
```

CONSTRUCT EQUATIONS WITH STANDARD ERRORS AND TEST STATISTICS

```
FIGREL1 =F2  =     .999*F1    + 1.000 D2
                   .150
                  6.676

FIGREL2 =F3  =     .749*F2    +  .671*F1    + 1.000 D3
                   .105          .180
                  7.103         3.736
```

This is the final solution for the modified structural regression model. Note that all structural regression coefficients postulated by the model are significant—their t values are all well outside the nonsignificance region (i.e., -2; $+2$). These results indicate that there are marked relationships between the Induction and Figural relations latent variables in both the junior and senior years. That is, the Induction latent variable (F_1) has marked explanatory power for individual differences in Figural relations at both assessment points (i.e., F_2 and F_3) because its regression coefficients, .999 and .671, are significant. As such, the relationship between junior- and senior-year Figural relations cannot be explained without considering the impact of their common predictor, Induction. It is clear, therefore, that in the presence of the Induction latent variable, Figural relations during the junior year is important for predicting individual differences in senior-year figural ability.

```
VARIANCES OF INDEPENDENT VARIABLES
----------------------------------
                  V                          F
                 ---                        ---
                          I  F1 -INDUCTN        25.165*I
                          I                      5.129 I
                          I                      4.907 I
                          I                            I



                E                          D
               ---                        ---
E1 - IND1           31.045*I  D2 -FIGREL1       39.881*I
                     3.880 I                     6.264 I
                     8.002 I                     6.366 I
                          I                            I
```

```
E2 - IND2                34.912*I  D3 -FIGREL2                39.366*I
                          5.052 I                              6.050 I
                          6.911 I                              6.506 I
                                I                                    I
E3 - IND3                24.322*I                                    I
                          3.060 I                                    I
                          7.948 I                                    I
                                I                                    I
E4 - FR11                19.667*I                                    I
                          3.352 I                                    I
                          5.868 I                                    I
                                I                                    I
E5 - FR12                27.709*I                                    I
                          3.533 I                                    I
                          7.842 I                                    I
                                I                                    I
E6 - FR13                28.541*I                                    I
                          3.463 I                                    I
                          8.242 I                                    I
                                I                                    I
E7 - FR21                29.401*I                                    I
                          4.279 I                                    I
                          6.871 I                                    I
                                I                                    I
E8 - FR22                31.342*I                                    I
                          3.952 I                                    I
                          7.930 I                                    I
                                I                                    I
E9 - FR23                22.501*I                                    I
                          3.280 I                                    I
                          6.860 I                                    I
                                I                                    I


COVARIANCES AMONG INDEPENDENT VARIABLES
-----------------------------------

                   E                          D
                   ---                        ---
E9 - FR23                12.264*I                              I
E6 - FR13                 2.460 I                              I
                          4.985 I                              I
                                I                              I
```

This output represents the variance of the Induction latent variable and the variances of the disturbance terms (D_2 and D_3) for the Figural relations latent variable. The covariance between the error terms of the fluid measures FR13 and FR23 are also reported in the section entitled COVARIANCES AMONG INDEPENDENT VARIABLES and, as expected, this value is found to be significant (see further discussion following).

```
COVARIANCES AMONG INDEPENDENT VARIABLES
-----------------------------------
```

	E		D
	---		---
E9 - FR23	.484*I		I
E6 - FR13	I		I
	I		I

The covariance between the error terms of the Figural relations measures (FR13 and FR23) is also provided as a correlation coefficient. As shown, despite the fact the correlation is only moderate in magnitude, it contributes to a substantial improvement in model. In fact, the improvement in fit can be easily judged by comparing the chi-square values for the two proposed models (i.e., the model with the correlated error terms versus the model without the correlated error terms). The difference in chi-square values of the two proposed models is found to be $\Delta T = 52.097 - 20.552 = 31.545$, with the difference in degrees of freedom, $\Delta df = 24 - 23 = 1$, and hence significant (cut-off value of the chi-square distribution with 1 df is 3.84 at significance level .05). Thus, it is evident that including the correlated error term leads to a considerable improvement in model fit.

TESTING FACTORIAL INVARIANCE ACROSS TIME

An important question frequently considered in studies that involve repeated measurements of latent variables concerns the invariance of the indicators used across time. A test of invariance addresses the basic question about the comparability of the measurement across time periods (in the example study it would be from the junior to senior years of high school). Thus, testing for invariance effectively focuses on whether the construct measured at repeated assessment occasions remains the same or whether it changes its structure. Because a necessary condition for measuring the same constructs across time is the identity of the factorial structure (i.e., the identity of the factor loadings), the term factorial invariance is commonly used. As it turns out, an examination of factorial invariance is possible by simply imposing an equality constraint on the factor structure and testing the resulting difference in the chi-square values for the two tested models for significance.

A test of factorial invariance in the example concerning mental ability involves examining whether the construct Figural relations retains its structure from the junior to senior high school year. This test can be accomplished by introducing an equality restriction on the factor loadings of the figural indicators across both assessment points. The equality restriction can be included in the EQS input file by adding a /CONSTRAINT section, and in the LISREL input file by adding EQuality lines as follows (recalling that the loading of the first indicator of the construct is already set equal to 1 for both assessment times):

/CONSTRAINTS
(V5,F2)=(V8,F3);
(V6,F2)=(V9,F3);

and

EQ LY(5, 2) LY(8, 3)
EQ LY(6, 2) LY(9,3)

The result of introducing this restriction leads to a chi-square value of T = 20.937, with df = 25 (recall that the model tested without the equality restriction resulted in a chi-square value of T = 20.552, with df = 23). The difference in chi-square values between the restricted model and the previous one is ΔT = 20.937 – 20.552 = 0.385, with Δdf = 2, and hence nonsignificant (the cut-off value of the chi-square distribution at significance level .05 is 5.99). This result indicates that the time-invariance restrictions imposed on the loadings of the Figural relations indicators are plausible. Because the restriction is found to be acceptable, it is retained in the model.

Next the LISREL output file only for the restricted model is presented and discussed. Comments are inserted at appropriate places and repetitive material is omitted (interested readers can easily very that the results of the LISREL and EQS programs are essentially the same by running the appropriate EQS input file).

PARAMETER SPECIFICATIONS

    LAMBDA-Y

	INDUCTN	FIGREL1	FIGREL2
IND1	0	0	0
IND2	1	0	0
IND3	2	0	0
FR11	0	0	0
FR12	0	3	0
FR13	0	4	0
FR21	0	0	0
FR22	0	0	3
FR23	0	0	4

As shown, all parameters constrained to be equal to one another are given the same number, whereas parameters that are fixed are not numbered. According to this section output, LISREL understood that this is a model with only four factor loading parameters to be estimated. The remaining parameters (including the correlated error term) are presented below for a total of 20 model parameters.

```
        BETA

              INDUCTN      FIGREL1      FIGREL2
              -------      -------      -------
INDUCTN          0            0            0
FIGREL1          5            0            0
FIGREL2          6            7            0


        PSI

              INDUCTN      FIGREL1      FIGREL2
              -------      -------      -------
                 8            9            0


        THETA-EPS

              IND1         IND2         IND3         FR11         FR12         FR13
              ----         ----         ----         ----         ----         ----
IND1            11
IND2             0           12
IND3             0            0           13
FR11             0            0            0           14
FR12             0            0            0            0           15
FR13             0            0            0            0            0           16
FR21             0            0            0            0            0            0
FR22             0            0            0            0            0            0
FR23             0            0            0            0            0           19


        THETA-EPS

              FR21         FR22         FR23
              ----         ----         ----
FR21            17
FR22             0           18
FR23             0            0           20
```

Next are the LISREL parameter estimates. Note in particular the restrained factor loadings that have identical estimates, standard errors, and *t* values.

```
LISREL ESTIMATES (MAXIMUM LIKELIHOOD)

        LAMBDA-Y

              INDUCTN      FIGREL1      FIGREL2
              -------      -------      -------
IND1           1.000          - -          - -
```

IND2	1.271	- -	- -
	(.157)		
	8.074		
IND3	.894	- -	- -
	(.116)		
	7.711		
FR11	- -	1.000	- -
FR12	- -	.879	- -
		(.040)	
		22.130	
FR13	- -	.855	- -
		(.041)	
		20.678	
FR21	- -	- -	1.000
FR22	- -	- -	.879
			(.040)
			22.130
FR23	- -	- -	.855
			(.041)
			20.678

BETA

	INDUCTN	FIGREL1	FIGREL2
	-------	-------	-------
INDUCTN	- -	- -	- -
FIGREL1	.992	- -	- -
	(.146)		
	6.776		
FIGREL2	.674	.758	- -
	(.180)	(.100)	
	3.747	7.617	

As can be seen, although the parameter estimates have changed slightly due to the imposed restrictions, the same substantive conclusions are warranted with respect to the structural regression coefficients examined in the earlier unconstrained model.

COVARIANCE MATRIX OF ETA

	INDUCTN	FIGREL1	FIGREL2
	-------	-------	-------
INDUCTN	25.179		
FIGREL1	24.965	64.179	
FIGREL2	35.891	65.484	113.456

PSI

INDUCTN	FIGREL1	FIGREL2
25.179	39.426	39.625
(5.130)	(5.806)	(5.927)
4.908	6.790	6.686

In the LISREL output presented in chapter 4 (for a confirmatory factor analysis model), the values in the COVARIANCE MATRIX OF ETA and those in the PSI matrix were identical. The results here, however, present very different values for the COVARIANCE MATRIX OF ETA and the PSI matrix. The reason that these values are different is that in this model explanatory relationships are assumed between the latent variables. As a result, the PSI matrix now contains both the variance of the first latent variable (INDUCTN) and the residual variances of the other two latent variables (FIGREL1 and FIGREL2). Thus, only the first element on its diagonal is identical to that of the COVARIANCE MATRIX OF ETA—the first latent factor is the common predictor for the other two and not regressed on any construct. Note that the structural residual variances are both significant (see the second and third diagonal elements of the matrix PSI), which reflects the fact that the latent relationship assumed in the model is not able to explain all individual variability in junior- and senior-year Figural relations.

SQUARED MULTIPLE CORRELATIONS FOR STRUCTURAL EQUATIONS

INDUCTN	FIGREL1	FIGREL2
- -	.386	.651

Based on these results, it appears that the model is the least successful in predicting junior-year Figural relations. Perhaps, a follow-up study should also consider other variables (beyond the Induction construct) that may contribute to a better explanation of individual differences in the Figural relations ability construct during junior high school year.

GOODNESS OF FIT STATISTICS

CHI-SQUARE WITH 25 DEGREES OF FREEDOM = 20.937 (P = 0.696)
ESTIMATED NON-CENTRALITY PARAMETER (NCP) = 0.0
90 PERCENT CONFIDENCE INTERVAL FOR NCP = (0.0 ; 10.044)

MINIMUM FIT FUNCTION VALUE = 0.0956
POPULATION DISCREPANCY FUNCTION VALUE (F0) = 0.0
90 PERCENT CONFIDENCE INTERVAL FOR F0 = (0.0 ; 0.0459)
ROOT MEAN SQUARE ERROR OF APPROXIMATION (RMSEA) = 0.0
90 PERCENT CONFIDENCE INTERVAL FOR RMSEA = (0.0 ; 0.0428)
P-VALUE FOR TEST OF CLOSE FIT (RMSEA < 0.05) = 0.977

```
        EXPECTED CROSS-VALIDATION INDEX (ECVI) = 0.278
90 PERCENT CONFIDENCE INTERVAL FOR ECVI = (0.297 ; 0.343)
           ECVI FOR SATURATED MODEL = 0.411
         ECVI FOR INDEPENDENCE MODEL = 5.458


CHI-SQUARE FOR INDEPENDENCE MODEL WITH 36 DEGREES OF FREEDOM = 1177.363
               INDEPENDENCE AIC = 1195.363
                    MODEL AIC = 60.937
                  SATURATED AIC = 90.000
               INDEPENDENCE CAIC = 1234.905
                   MODEL CAIC = 148.809
                 SATURATED CAIC = 287.713


        ROOT MEAN SQUARE RESIDUAL (RMR) = 1.462
               STANDARDIZED RMR = 0.0182
          GOODNESS OF FIT INDEX (GFI) = 0.980
    ADJUSTED GOODNESS OF FIT INDEX (AGFI) = 0.964
   PARSIMONY GOODNESS OF FIT INDEX (PGFI) = 0.544


           NORMED FIT INDEX (NFI) = 0.982
         NON-NORMED FIT INDEX (NNFI) = 1.005
      PARSIMONY NORMED FIT INDEX (PNFI) = 0.682
        COMPARATIVE FIT INDEX (CFI) = 1.000
        INCREMENTAL FIT INDEX (IFI) = 1.004
         RELATIVE FIT INDEX (RFI) = 0.974


               CRITICAL N (CN) = 464.534
```

The goodness-of-fit indices all point to the model as a reasonable approximation of the data. In particular, note the low magnitude of the RMSEA index (well below the proposed threshold of .05). Moreover, the left endpoints of the confidence interval of the RMSEA index and that of the noncentrality parameter are 0, which is another sign of a satisfactory model fit.

```
SUMMARY STATISTICS FOR FITTED RESIDUALS
  SMALLEST FITTED RESIDUAL =    -3.756
   MEDIAN FITTED RESIDUAL =      .000
  LARGEST FITTED RESIDUAL =     4.197

STEMLEAF PLOT
 - 3|8
 - 2|876
 - 1|75544321
 - 0|966544431000
   0|2334566689
   1|000233477
   2|3
   3|
   4|2
```

```
SUMMARY STATISTICS FOR STANDARDIZED RESIDUALS
SMALLEST STANDARDIZED RESIDUAL =      -1.556
  MEDIAN STANDARDIZED RESIDUAL =        .000
 LARGEST STANDARDIZED RESIDUAL =       1.344

STEMLEAF PLOT
 - 1|6
 - 1|0
 - 0|97755
 - 0|44444332222221000
   0|1111222333344
   0|556667
   1|03
```

None of the standardized residuals is very large, which is another indication that the model also fits well locally (i.e., beyond being a good overall means of data description).

```
MODIFICATION INDICES AND EXPECTED CHANGE

        MODIFICATION INDICES FOR LAMBDA-Y
```

	INDUCTN	FIGREL1	FIGREL2
IND1	- -	.010	.206
IND2	- -	.241	.374
IND3	- -	.366	.027
FR11	.011	.060	.002
FR12	.415	.181	.268
FR13	.479	.360	.259
FR21	.203	.035	.060
FR22	.131	.180	.181
FR23	.559	.299	.360

```
             EXPECTED CHANGE FOR LAMBDA-Y
```

	INDUCTN	FIGREL1	FIGREL2
IND1	- -	.009	-.036
IND2	- -	.057	.060
IND3	- -	-.051	-.012
FR11	.011	.018	.002
FR12	.067	.019	.024
FR13	-.063	-.026	-.021
FR21	-.058	-.015	-.018
FR22	-.045	-.032	-.012
FR23	.079	.035	.010

```
NO NON-ZERO MODIFICATION INDICES FOR BETA

NO NON-ZERO MODIFICATION INDICES FOR PSI
```

MODIFICATION INDICES FOR THETA-EPS

	IND1	IND2	IND3	FR11	FR12	FR13
IND1	- -					
IND2	.143	- -				
IND3	.440	.081	- -			
FR11	.368	.050	.818	- -		
FR12	.049	.001	.480	.044	- -	
FR13	1.644	.128	4.902	.025	.127	- -
FR21	.799	.106	.001	6.350	7.813	.006
FR22	1.635	.001	1.863	.806	.583	.009
FR23	.440	.003	1.681	3.262	4.695	- -

MODIFICATION INDICES FOR THETA-EPS

	FR21	FR22	FR23
FR21	- -		
FR22	.016	- -	
FR23	.001	.008	- -

EXPECTED CHANGE FOR THETA-EPS

	IND1	IND2	IND3	FR11	FR12	FR13
IND1	- -					
IND2	-1.605	- -				
IND3	1.976	-1.080	- -			
FR11	-1.429	-.592	1.891	- -		
FR12	-.548	.106	1.520	.680	- -	
FR13	2.882	.900	-4.418	.466	-.931	- -
FR21	-2.424	.987	.079	6.580	-7.397	-.208
FR22	3.371	-.104	-3.194	-2.213	1.970	.234
FR23	-1.438	-.143	2.496	-3.973	4.685	- -

EXPECTED CHANGE FOR THETA-EPS

	FR21	FR22	FR23
FR21	- -		
FR22	.529	- -	
FR23	-.097	-.308	- -

MAXIMUM MODIFICATION INDEX IS   7.81 FOR ELEMENT (7, 5) OF THETA-EPS

Although there is a modification index that is larger than 5 in this output—see element (7,5) of the TE matrix—adding this parameter to the proposed model cannot be theoretically justified because there does not appear to be a good substantive reason for the two error terms to be correlated. In addition, because the model fit is already acceptable, no further change need be made to the proposed model. One can therefore consider the model with the factorial invariance across time to be an acceptable means of data description.

CHAPTER SIX

Latent Change Analysis

WHAT IS LATENT CHANGE ANALYSIS?

Repeated measurements are used quite often in research studies to examine changing processes in individuals or groups. One reason for their popularity is that repeated measurements allow researchers to investigate individual (intraindividual) development across time as well as between-individual (interindividual) differences and similarities in change patterns. For example, educational researchers may be interested in examining particular patterns of growth (or decline) in ability exhibited by a group of subjects in repeatedly administered measures following some treatment program. The researchers may also be interested in comparing the rates of change in these variables across several student populations. Finally, the researchers may be interested in studying the correlates and predictors of growth (or decline) in ability over time in an effort to determine which students exhibited the fastest improvement in ability.

A powerful methodology for addressing these kinds of questions exists within the traditional analysis of variance (ANOVA) and analysis of covariance (ANCOVA) frameworks. Unfortunately, some of the assumptions needed to use this methodology can often be untenable. In particular, assumptions concerning the homogeneity of the variance–covariance matrix across the levels of the between-subjects factors, or the assumption of sphericity (implying the same patterns of correlation for repeated assessments) can be problematic (e.g., Marcoulides & Hershberger, 1997; Tabachnick & Fidell, 1999). Similarly, when studying correlates and predictors of growth or decline via ANCOVA, assumptions concerning the use of perfectly measured covariate(s) and regression homogeneity may also be questionable (Huitema, 1980). As it turns out, the SEM methodology offers a useful alternative framework, latent change analysis (LCA). LCA has been developed over the past few decades and popularized in the social and behavioral sciences as a general means for studying latent growth or decline, as well as their correlates and predictors. In many ways, the traditional repeated measures analysis of variance models can be considered to be special cases of a more general LCA model (Meredith and Tisak, 1990). However, the LCA model provides more flexibility in the measurement of change than those traditional analysis of variance models. The LCA model also resembles the confirmatory factor analysis model discussed in chapter 4 with one exception. Because the LCA model uses data obtained from repeated measurements, the latent variables are interpreted as chronometric variables representing individual differences over time.

The term latent change analysis (LCA) is used in this book as a general category for a large number of possible models that can be used in repeated-assessment research studies. Many of these models have been referred to in the literature under various related names, such as latent growth curve models, latent curve analysis models, or just growth curve models. We prefer to refer to the models as latent change analysis models to emphasize that the models are equally applicable to all cases in which one is interested in studying growth or decline, even those with a more complex pattern of change such as growth followed by decline or vice versa. Due to the introductory nature of this book, only two LCA models are introduced: (a) the one-factor model and (b) the two-factor (or LS) model. For more extensive and advanced treatments of the subject, several excellent sources are available (e.g., Duncan, Duncan, Strycker, Li, and Alpert, 1999; McArdle, 1998; Meredith & Tisak, 1990; Willett & Sayer, 1996).

SIMPLE ONE-FACTOR LATENT
CHANGE ANALYSIS MODEL

To introduce the reader to latent change modeling, the discussion of LCA begins with a simple one-factor model. In this simple model, it is assumed that a set of $k = 4$ repeated measurements of cognitive ability give rise to a covariance matrix and means that can be explained in terms of a single latent variable (see Fig. 12 and details in the following). McArdle (1988) has termed the one-factor LCA model a curve model because the latent variable can be interpreted as a time factor that governs the intraindividual latent change curves (see also McArdle & Epstein, 1987). Alternatively, the time factor can be interpreted as an initial true status of the underlying ability that is being repeatedly measured. As indicated later, the loadings

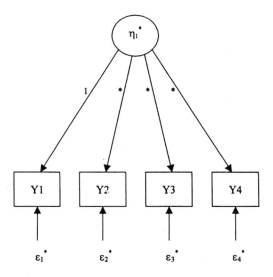

FIG. 12.   Example one-factor LCA model.

of the repeated measures on this factor can be interpreted as rates of mean change in the studied ability. Meredith and Tisak (1990) chose to term this model a monotonic stability model because they felt that, although significant changes in mean levels may occur in the studied ability, the rank order of the observations stays the same over repeated measures (Duncan et al., 1999).

Analysis of Mean Structures

One important aspect in which the LCA models differ from all the preceding ones discussed in this book is that the variable means and their development over time are taken into account. This is achieved by what is referred to as mean structure analysis (MSA). A MSA includes in the analysis not only the covariance matrix of the repeated measures, but also the observed variable means. In order to achieve this, a model must be fitted to the covariance–mean matrix, rather than only to the covariance matrix that has been discussed up to this point. The covariance–mean matrix results after the observed variable means are added as a last row and column to the covariance matrix (i.e., the covariance–mean matrix is the covariance matrix augmented by the observed variable means).

Why Is It Necessary to Include Variable Means into an Analysis of Change? The inclusion of observed variable means into the analysis complies with the conceptual notions of the classical approaches to studying change. According to these notions, development in the means of the observed variables under investigation is of special importance. Indeed, one cannot imagine repeated measures ANOVA that would exclude the information about temporal change contained in the means. Because the use of LCA has the same goal of studying development over time, it seems only natural to include the variable means in the analysis. If the model were fitted only to the covariance matrix, however, one would omit the observed variable means from the analysis. The reason is that any covariance coefficient is defined in terms of the sum of cross-products around the means (e.g., Hays, 1994). That is, any covariance (as well as correlation) disregards the means of the two involved variables, and as a result the observed means and their development over time are inconsequential for the covariance matrix. Hence, different patterns of change over time (e.g., increase over time; decline; or growth followed by decline, or vice versa) can be equally consistent with a given covariance matrix resulting in a repeated measure context. It follows, therefore, that to fit a model only to the covariance matrix is tantamount to being wasteful with information about the development of the studied phenomenon over time, which is contained in the observed variable means. Therefore, in analyses of change the observed variable means are always included because, without them, one cannot achieve the goal of examining growth or decline.

As a result of this inclusion, more data points are obtained to which the model is fitted. Therefore, in addition to the elements of the covariance matrix, there are also as many means to count as there are observed variables. Thus, with $k = 4$ repeated measurements assessments, there are $k(k + 1)/2 = 4(5)/2 = 10$ nonredundant elements of the covariance matrix plus $k = 4$ observed means that the model must also emulate. Hence, in a mean structure analysis of $k = 4$ variables there are, altogether, $q = k(k + 1)/2 + k = 4(5)/2 + 4 = 14$ pieces of empirical information to which the model must be fit. Of course, from this number q it is still necessary to subtract the number of model parameters in order to obtain the model degrees of freedom. As shown later, conducting a MSA makes necessary the introduction of additional model parameters concerning the structure of the variable means. This issue will be discussed further after details concerning LCA models are presented.

How Is a Model Fitted to the Covariance–Mean Matrix?   SEM accomplishes the inclusion of observed variable means in the analysis (i.e., achieves fitting a model to the sample covariance–mean matrix), by extending the fit function with a special term (see discussion of the fit function in chap. 1). The special term represents a weighted sum of the squared differences between the observed means and those reproduced by the model, with the weights being the corresponding elements of the model-reproduced covariance matrix Σ (for the maximum likelihood method used throughout this book). Just as the model has certain implications for the variances and covariances of the analyzed variables, it also has consequences for the variable means. These consequences can easily be worked out by noting that the mean of any linear combination of variables—whether observed or latent—is simply the same linear combination of their means. In order to understand and conceptualize the LCA model implications and the reproduced covariance–mean matrix, a fifth rule is now added to those presented earlier in chapter 1.

Law 5 of Variable Means. The relationship between means of linear combination of variables and means of its constituents is formulated as a specific law as follows (and added to the four laws of variable variances and covariances discussed in chapter 1):

Law 5. For any two variables X and Y and any given constants a and b,

$$M(aX + bY) = aM(X) + bM(Y),$$

where $M(\cdot)$ denotes the mean. It is important to note that the validity of this law follows directly from the additive property of the mean (e.g., see Hays, 1994.). Of course, Law 5 can obviously be generalized to any number of variables—the mean of a linear combination of variables is the same linear combination of their means.

Once advised to perform a mean structure analysis, most structural equation modeling programs will automatically augment the fit function by this special term, evaluating model fit with regard to means. A minimization of the augmented fit function thus implies that when fitting a model to the mean structure one is seeking estimates of its parameters that render the reproduced covariance–mean matrix as close as possible to the observed covariance–mean matrix. Therefore, at the final computed solution, the observed variances, covariances, and means are emulated as well as possible by the fitted model.

The One-Factor Latent Change Analysis Model

Returning to the specific one-factor model presented in Fig. 12, note that the model path diagram is presented in LISREL notation and includes the four observed variables (Y's), each indexed according to the measurement occasion. The model also proposes that each observed variable Y loads on a single latent construct η_1, which represents the time factor underlying their change as indicated previously, and involves $k = 4$ successive measurements.

To determine the parameters of the model presented in Fig. 12 (designated by asterisks), one must follow the six rules outlined in chapter 1. According to Rule 1, all error-term variances and the variance of the factor η_1 are parameters, and, according to Rule 3, all factor loadings are parameters. Rule 2 is not applicable to this model because there are no variable covariances (i.e., there are no two-way arrows in Fig. 12). Rule 6 requires that the scale of the latent variable be fixed, which can be accomplished by fixing either the variance of η_1 or one of the factor loadings to a value of 1. Here, fix the scale of the latent variable by fixing the first measurement loading to a value of 1 (i.e., set the loading of Y_1 equal to 1). Using this approach effectively sets the metric of the first measurement as a baseline against which the subsequent change in the repeated assessments is estimated. This facilitates parameterizing the rates of change in means over time into the factor loadings (see the following discussion).

Now to demonstrate an analysis on the basis of this proposed model, consider the following data based partly on a study of cognitive intervention by Baltes, Dittmann-Kohli, and Kliegl (1986). The goal of the original study was to examine the extent of reserve capacity of older adults in test performance on repeatedly presented fluid-intelligence measures. As part of the study, $N = 161$ older adults were administered tutor-guided training in test-relevant skills that focused in part on a specific component of fluid intelligence, namely Induction (a discussion concerning the construct of fluid intelligence can be found in chap. 5). Four repeated assessments were carried out using Thurstone's Standard Induction test: (a) before the training program was administered (Y_1), and then (b) 1 week after (Y_2), (c) 1 month after (Y_3), and (d) 6 months after completion of training (Y_4). The covariance matrix and the observed variable means for this example one-factor model are provided in Table 3. The means of the four repeated assessments are presented in the last column of Table 3 to em-

TABLE 3
Covariance Matrix and Means of the Four Consecutive
Administrations of Thurstone's Standard Induction Test

	THU_IND1	THU_IND2	THU_IND3	THU_IND4	Means
THU_IND1	307.46				37.48
THU_IND2	296.52	377.21			53.30
THU_IND3	295.02	365.10	392.47		54.82
THU_IND4	291.02	355.88	358.25	376.84	52.63

Note. THU_INDi = Thurstone's Standard Induction test at ith assessment ($i = 1, 2, 3, 4$).

phasize that they are essential for achieving the goals of latent change analysis.

LISREL AND EQS INPUT FILES

The input files for the LISREL and EQS programs must indicate that the model should be fitted to both the covariance matrix and the means. Accomplishing this necessitates a slight modification of the input files that are used when fitting a model only to the covariance matrix. The modification actually follows from an observation about model-reproduced means discussed in the previous section: Just as there is a model-reproduced covariance matrix Σ, so are there also model-reproduced means that are obtained by applying Law 5 to the model-definition equations. For example, consider the following equation for Y_2 in Fig. 12:

$$Y_2 = \lambda_2 \eta_1 + \varepsilon_2, \qquad (12)$$

where λ_2 is its factor loading on η_1 and ε_2 is the corresponding error term. Applying Law 5 to Equation 12, the following mean value for Y_2 is obtained:

$$M(Y_2) = \lambda_2 M(\eta_1) + M(\varepsilon_2), \qquad (13)$$

but because $M(\varepsilon_2) = 0$ (for further discussion, see the discussion of error means in a structural model in chap. 1), the equation becomes

$$M(Y_2) = \lambda_2 M(\eta_1). \qquad (14)$$

In other words, the one-factor model under consideration reproduces the mean of Y_2 as the product of the latent variable mean and the measure's factor loading. The same relationship applies to the means of all observed variables (keeping in mind that $\lambda_1 = 1$ due to the earlier decision to set the latent metric by fixing that loading).

Hence, in order to analyze latent change, one must analyze the mean structure of the observed variables (i.e., fit the model to the covariance–mean matrix), and it is meaningful to also introduce the latent variable mean $M(\eta_1)$ as a parameter. Otherwise, if no latent variable mean is introduced one is assuming that its value is effectively equal to zero. Obviously, this would lead to all the observed variable means being reproduced as zero, as implied from Equations 12 and 14, and almost certainly would lead to a misfit of the proposed model. Thus, in the model under consideration one needs to include (at least) the latent mean as a separate parameter in order to ensure that the model is given a chance to fit the data as well as it possibly can.[1]

Continuing with the earlier discussion of parameters, the model under consideration in Fig. 12 will now have a total of nine parameters (i.e., four error variances, three factor loadings, one latent variance, and the latent variable mean). Because this is a LCA model, it is fit to the covariance–mean matrix as discussed previously. Therefore, with four repeated assessments (corresponding to the observed variables), it is fit to a total of $4(5)/2 + 4 = 14$ data points (i.e., the 10 nonredundant elements of the covariance matrix plus the four manifest means). The model thus has $14 - 9 = 5$ degrees of freedom.

Equation 14 also permits a direct interpretation of the factor loadings as rates of mean change. For example, if Law 5 is applied to the equation for the first occasion of measurement, $Y_1 = \eta_1 + \varepsilon_1$, this yields the mean value, $M(Y_1) = M(\eta_1)$. If one inserts the value $M(Y_1)$ for $M(\eta_1)$ in Equation 14 and expresses the equation in terms of λ_2, one obtains $\lambda_2 = M(Y_2)/M(Y_1)$. This result implies that the value of the factor loading λ_2, under the considered model, corresponds to the ratio of the means for that measurement occasion and the first assessment (used to fix the scale of the latent variable). It turns out that this idea can actually be generalized to all the remaining factor loadings. Thus, the value of any factor loading included in the model is obtained as a ratio of two means (for that measurement occasion and the first assessment).

LISREL Input File

The LISREL input file for conducting a LCA requires that the latent variable mean be introduced. Accordingly, latent variable means are referred to as part of an A matrix (the Greek letter *alpha*), denoted as AL in the LISREL syntax, and the same principles discussed in chapter 2 apply when referring to its elements. It is important to note that A is actually a row vector—a special matrix that has only one row, but as many columns as there are latent variables in the model. Thus, the elements of the ALpha matrix in this example are AL(1), or simply AL(1, 1)—recall that there is only one latent variable mean. The default setting for the ALpha matrix in LISREL is that it is fixed (i.e., it consists only of elements fixed at 0), unless the program is advised otherwise. Thus, once ALpha is mentioned in the model-definition line, LISREL is prepared for any instructions concerning latent means.

---

[1]There is a more general treatment of mean structure analysis models with additional mean structure parameters, the mean intercepts, which we will not present in this text. We refer the reader to Jöreskog & Sörbom (1993b, 1999) or Bentler (1995, 2000).

The LISREL input file here is constructed following the principles outlined in chapter 2 and includes the definition of the ALpha matrix.

```
ONE-FACTOR LATENT CHANGE MODEL
DA NI=4 NO=161
CM
307.46
296.52   377.21
295.02   365.10   392.47
291.02   355.88   358.25   376.84
ME
37.48   53.30   54.82   52.63
LA
THU_IND1 THU_IND2 THU_IND3 THU_IND4
MO NY=4 NE=1 AL=FR LY=FU, FR
LE
TIME_FAC
FI LY(1, 1)
VA 1 LY(1, 1)
OU
```

There are several items in the LISREL input file that should be mentioned. First, there is a new line immediately following the covariance matrix that corresponds to the four means of the observed variables. This is
the reason the command line uses the keyword ME (for MEans). The observed variables are subsequently labeled according to the four consecutive assessments using Thurstone's Induction test (i.e., THU_IND1 to
THU_IND4). The MOdel definition line also includes the new command,
AL=FR. This command tells the LISREL program that the latent means are
to be considered free model parameters by effectively freeing all elements
of ALpha. Of course, because the model being fitted contains only one factor, ALpha has only one element, and hence this latent mean is hereby declared a free parameter. Then, for simplicity, the matrix LY is declared to
be full of free parameters. Because LY has here only one column containing all four factor loadings (as this model has just one latent variable), this
in effect says that all of them are free model parameters (of course, the
first loading is later fixed to 1 to set the scale metric). Note that the PS matrix, here consisting of only one latent variance, has not been mentioned.
This is because PS=DI, FR (i.e., PS having free elements along its diagonal)
is the default option for it, which is exactly what the proposed model entails. Similarly, the TE matrix containing all error variances has not been
mentioned, because TE=DI, FR is also the default and just what is needed
to specify it in this example. Finally, below the MOdel definition line, a la-

bel TIME_FAC (for TIME FACtor) is provided for the single latent variable in the model.

EQS Input File

The EQS input file for conducting a LCA requires that the latent variable mean be introduced using a special auxiliary variable. The name of the variable is V999. According to Bentler (1995), the variable designation V999 was chosen so that it "will always be in the same position, the last, relative to the measured variables in the input file" (p. 166). V999 can be thought of as a dummy variable with two features: (a) it takes on the value of 1 for all subjects in the sample, and (b) it is added internally by the program once it is designated in the input file. Subsequently, by regressing the latent variable on this dummy variable, one obtains in the resulting slope estimate the value of the latent mean. Indeed, this is equivalent to using a simple regression model in which the latent variable is the dependent variable, V999 is the single predictor, and no intercept term is included; then the estimate of the slope of V999 will obviously be the latent mean. This is, alternatively, equivalent to using a simple regression model with an intercept only, which will then evidently be estimated at the mean of the dependent variable—in this case the latent variable mean. Of course, regressing the latent variable on the variable V999 implies that a residual term is to be associated with the factor. Recall the discussion in chapter 1 in which it was emphasized that in general any dependent variable should be associated with a residual (disturbance) term (which in EQS is denoted by D). This requirement leads to an additional model equation for the latent mean in the model (i.e., an added equation for the single factor in the model in Fig. 12). Thus, the equation $F_1 = *V999 + D_1$ is used in the EQS input file to regress the factor on the dummy variable V999, and as a result the residual term D_1 receives the variance of the latent variable as a model parameter. This is due to two reasons: (a) D_1 is formally an independent variable of the model and hence its variance is a model parameter, and (b) applying Law 4 for variances on the equation (see chap. 1), one can see that its variance equals that of the latent factor. In actuality, reason b follows from the fact that D_1 is assumed to be unrelated to F_1 and the fact that V999 has no variance because it is a dummy constant variable.

Because the input file being created is for the analysis of a LCA model, the EQS program must also be provided with data to fit the model to a covariance–mean matrix. In order to add the means of the observed variables to the input file, the command line /MEANS below the covariance matrix command line must be used (and in the same variable order as the covariance matrix). Once this is done, an analysis of the mean structure in the EQS program is invoked by using the keyword ANALYSIS=MOMENTS in the specification section (see also the discussion in chap. 2). The use of the keyword ANALYSIS=MOMENTS signals to the EQS program that both the first-order and second-order moments of the observed variables are to be analyzed (recall from introductory statistics that means are first-order moments, and their variances and covariances are second-order moments). Thus, the following EQS input file is created:

```
/TITLE
ONE-FACTOR LATENT CHANGE ANALYSIS MODEL;
/SPECIFICATIONS
VARIABLES=4; CASES=161; ANALYSIS=MOMENTS;
/LABELS
V1=THU_IND1; V2=THU_IND2; V3=THU_IND3; V4=THU_IND4;
F1=TIME_FAC;
/EQUATIONS
V1=  F1+E1;
V2=*F1+E2;
V3=*F1+E3;
V4=*F1+E4;
F1=*V999+D1;
/VARIANCES
D1=*; E1 TO E4=*;
/MATRIX
307.46
296.52   377.21
295.02   365.10   392.47
291.02   355.88   358.25   376.84
/MEANS
37.48   53.30   54.82   52.63
/END;
```

MODELING RESULTS

LISREL Program Results

The output produced by the LISREL input file created in the previous section is presented here. As before, comments are inserted at appropriate places to clarify portions of the output and redundant or irrelevant sections and repetitive page titles are omitted. The information on the goodness of fit of the model is presented before that on estimates, so that an interpretation of the estimates is only rendered after model tenability is ensured.

```
PARAMETER SPECIFICATIONS

        LAMBDA-Y

                TIME_FAC
                -------
THU_IND1              0
THU_IND2              1
THU_IND3              2
THU_IND4              3


        PSI

                TIME_FAC
                -------
                     4

        THETA-EPS

                THU_IND1        THU_IND2        THU_IND3        THU_IND4
                -------         -------         -------         -------
                     5              6               7               8
        ALPHA
                TIME_FAC
                -------
                     9
```

From this section, it is assured that the number and exact location of the nine model parameters—including the latent mean parameter in the matrix ALpha—have been communicated to the LISREL program.

```
                    GOODNESS OF FIT STATISTICS

        CHI-SQUARE WITH 5 DEGREES OF FREEDOM = 10.86 (P = 0.054)
              ESTIMATED NON-CENTRALITY PARAMETER (NCP) = 5.86
        90 PERCENT CONFIDENCE INTERVAL FOR NCP = (0.0 ; 19.46)

                  MINIMUM FIT FUNCTION VALUE = 0.068
          POPULATION DISCREPANCY FUNCTION VALUE (F0) = 0.037
          90 PERCENT CONFIDENCE INTERVAL FOR F0 = (0.0 ; 0.12)
        ROOT MEAN SQUARE ERROR OF APPROXIMATION (RMSEA) = 0.086
        90 PERCENT CONFIDENCE INTERVAL FOR RMSEA = (0.0 ; 0.16)
          P-VALUE FOR TEST OF CLOSE FIT (RMSEA < 0.05) = 0.17

              EXPECTED CROSS-VALIDATION INDEX (ECVI) = 0.18
        90 PERCENT CONFIDENCE INTERVAL FOR ECVI = (0.12 ; 0.24)
                    ECVI FOR SATURATED MODEL = 0.13
                  ECVI FOR INDEPENDENCE MODEL = 6.17

    CHI-SQUARE FOR INDEPENDENCE MODEL WITH 6 DEGREES OF FREEDOM = 979.11
                     INDEPENDENCE AIC = 987.11
                        MODEL AIC = 28.86
                      SATURATED AIC = 20.00
```

```
                    INDEPENDENCE CAIC = 1003.43
                         MODEL CAIC = 65.59
                     SATURATED CAIC = 60.81


            ROOT MEAN SQUARE RESIDUAL (RMR) = 25.07
                      STANDARDIZED RMR = 0.086
                 GOODNESS OF FIT INDEX (GFI) = 0.97
        ADJUSTED GOODNESS OF FIT INDEX (AGFI) = 0.94
      PARSIMONY GOODNESS OF FIT INDEX (PGFI) = 0.49


                   NORMED FIT INDEX (NFI) = 0.99
               NON-NORMED FIT INDEX (NNFI) = 0.99
           PARSIMONY NORMED FIT INDEX (PNFI) = 0.82
              COMPARATIVE FIT INDEX (CFI) = 0.99
              INCREMENTAL FIT INDEX (IFI) = 0.99
                 RELATIVE FIT INDEX (RFI) = 0.99


                    CRITICAL N (CN) = 223.29
```

Although not spectacular, all of the goodness-of-fit indices indicate an acceptable fit of the model for purposes of further discussion. And, although not displayed, the model residuals also do not leave a very different impression concerning model fit. One can therefore consider this model to be a reasonable approximation to the analyzed data and continue the discussion of the output, focusing on its parameter estimates. Note that the degrees of freedom for this model are 5 because this LCA model is fitted to the covariance–mean matrix rather than only to the covariance matrix.

```
LISREL ESTIMATES (MAXIMUM LIKELIHOOD)


        LAMBDA-Y


            TIME_FAC
            --------
THU_IND1      1.00

THU_IND2      1.40
             (.02)
             57.51

THU_IND3      1.43
             (.03)
             55.68

THU_IND4      1.38
             (.03)
             55.04
```

```
COVARIANCE MATRIX OF ETA

     TIME_FAC
     --------
      185.75

PSI

     TIME_FAC
     --------
      185.75
     (21.98)
        8.45

THETA-EPS

     THU_IND1    THU_IND2    THU_IND3    THU_IND4
     --------    --------    --------    --------
       71.99       13.96       25.56       27.43
      (8.44)      (3.07)      (4.05)      (4.08)
        8.53        4.55        6.32        6.72

SQUARED MULTIPLE CORRELATIONS FOR Y - VARIABLES

     THU_IND1    THU_IND2    THU_IND3    THU_IND4
     --------    --------    --------    --------
         .72         .96         .94         .93

ALPHA

     TIME_FAC
     --------
       38.14
      (1.25)
       30.50
```

Based on an evaluation of the factor loading estimates of the LY matrix, there seems to be evidence of mean growth along the studied dimension at all the posttests (Y_2 to Y_4) compared to the pretest (Y_1). For example, this can be seen by comparing the confidence intervals (CI) of each posttest factor loading with that of the pretest (i.e., the factor loading fixed at 1). Adding twice the standard errors to and subtracting twice the standard errors from each of the posttest loadings results in the 95% CI: (1.36; 1.44), (1.37; 1.49), and (1.32; 1.44). Because none of these includes the value of the factor loading at pretest (i.e., the value 1), it is suggested that each of the posttest loadings is markedly higher than the pretest loading. Given the interpretation that factor loadings are ratios of mean values (see the discussion in the previous section), it appears that at each posttest there is considerable growth in performance compared to the pretest. Take special note, however, that using this confidence-interval approach is not equivalent to formal hypothesis testing. Specifically, it is possible that conclusions arrived at with this method may not be confirmed by a formal hypothesis test, especially when more than two parameters are simultaneously compared. It is therefore recommended that this confidence-interval approach be used only as a rough explorative method to examine parameter relationships informally.

The three CIs considered here also overlap to a considerable degree, suggesting that the posttest factor loadings may be fairly similar in magnitude. To examine this suggestion formally via a statistical test, one can impose a restriction on the proposed LCA model and add the following LISREL EQuality command line:

$$\text{EQ LY(2, 1) LY(3, 1) LY(4, 1)}$$

As it turns out, this equality constraint leads to a substantial decrement in model fit and is associated with a chi-square value of $T = 25.68$ with $df = 7$. The difference in chi-square values compared to the originally proposed model is $\Delta T = 15.822$ with $\Delta df = 2$, and hence significant (the cut-off value for the pertinent chi-square distribution with 2 degrees of freedom is 5.99 at significance level .05). This finding indicates that the rate of change in means is not the same over all the posttests when compared to the pretest.

Because in the preceding model the factor loading for the third assessment occasion was highest and given the lack of equality for factor loadings observed previously, it seems reasonable to test if an identity exists only between the other two posttest occasions (i.e., between the second and fourth assessment occasions). Introducing this restriction, by deleting LY(3, 1) in the EQuality command line, leads to a chi-square value of $T = 13.00$ with $df = 6$. This value is somewhat better than that of the last restricted model; the chi-square difference is $\Delta T = 25.68 - 13.00 = 12.65$ with $\Delta df = 7 - 6 = 1$, but it is still significant (using the chi-square distribution cut-off value of 3.84 at the .05 level of significance). The finding indicates that the second posttest loading is not identical to that of the first and last posttest occasions. On the other hand, because this model is nested in the first model that was tested (i.e., the model with no restrictions at all on the factor loadings), one can also examine their chi-square differences. Now the difference in their chi-square values is equal to $\Delta T = 13.00 - 10.86 = 2.14$ with $\Delta df = 6 - 5 = 1$ and is nonsignificant. This result indicates that the factor loadings for the second and fourth assessment occasions are equal to one another, and the factor loading of the third assessment occasion is the highest of all four loadings. In other words, it has been determined that the mean performance at the third assessment occasion is the highest, that the mean performance at the second and four assessment occasions are similar to one another, and that all posttest means are markedly higher than that observed at the pretest. All these results suggest that there has indeed been growth in test perfor-

mance as a result of the cognitive intervention. At any subsequent assessment, subjects performed, on average, markedly better than they did at
baseline measurement (pretest) before receiving the training program.

EQS Program Results

Only the EQS output for the last fitted model, in which the factor loadings
of the second and fourth assessment occasion were set equal, are presented here. (Interested readers can easily verify that the results of the
LISREL and EQS programs are identical by running the appropriate
LISREL input file.)

```
PARAMETER ESTIMATES APPEAR IN ORDER,
NO SPECIAL PROBLEMS WERE ENCOUNTERED DURING OPTIMIZATION.

ALL EQUALITY CONSTRAINTS WERE CORRECTLY IMPOSED
```

As indicated in previous chapters, this message is a reassurance that the
program has not encountered problems stemming from lack of identification or other numerical difficulties and that the model is technically
sound.

```
RESIDUAL COVARIANCE/MEAN MATRIX (S-SIGMA):
```

		THU_IND1	THU_IND2	THU_IND3	THU_IND4	V999
		V 1	V 2	V 3	V 4	V999
THU_IND1 V	1	49.773				
THU_IND2 V	2	38.180	3.408			
THU_IND3 V	3	28.761	-5.330	-14.699		
THU_IND4 V	4	32.680	-3.532	-12.180	-10.083	
V999	V999	-0.658	0.241	0.134	-0.429	0.000

```
                   AVERAGE ABSOLUTE COVARIANCE RESIDUALS     =        13.3391
       AVERAGE OFF-DIAGONAL ABSOLUTE COVARIANCE RESIDUALS    =        12.2124


STANDARDIZED RESIDUAL MATRIX:
```

		THU_IND1	THU_IND2	THU_IND3	THU_IND4	V999
		V 1	V 2	V 3	V 4	V999
THU_IND1 V	1	0.162				
THU_IND2 V	2	0.112	0.009			
THU_IND3 V	3	0.083	-0.014	-0.037		
THU_IND4 V	4	0.096	-0.009	-0.032	-0.027	
V999	V999	-0.038	0.012	0.007	-0.022	0.000

```
                   AVERAGE ABSOLUTE STANDARDIZED RESIDUALS     =        0.0440
       AVERAGE OFF-DIAGONAL ABSOLUTE STANDARDIZED RESIDUALS    =        0.0425
```

LARGEST STANDARDIZED RESIDUALS:

```
 V  1,V  1    V  2,V  1    V  4,V  1    V  3,V  1    V999,V  1
    0.162        0.112        0.096        0.083       -0.038

 V  3,V  3    V  4,V  3    V  4,V  4    V999,V  4    V  3,V  2
   -0.037       -0.032       -0.027       -0.022       -0.014

 V999,V  2    V  4,V  2    V  2,V  2    V999,V  3    V999,V999
    0.012       -0.009        0.009        0.007        0.000
```

DISTRIBUTION OF STANDARDIZED RESIDUALS

```
     - - - - - - - - - - - - - - - - - - - - - - - - - - -
      !                               !
  20- !                               -
      !                               !
      !                               !
      !                               !
      !                               !           RANGE      FREQ  PERCENT
  15- !                               -
      !                               !   1  -0.5  -  --        0   0.00%
      !                               !   2  -0.4  - -0.5       0   0.00%
      !                               !   3  -0.3  - -0.4       0   0.00%
      !                               !   4  -0.2  - -0.3       0   0.00%
  10-                                 -   5  -0.1  - -0.2       0   0.00%
      !                               !   6   0.0  - -0.1       8  53.33%
      !              *                !   7   0.1  -  0.0       5  33.33%
      !              *                !   8   0.2  -  0.1       2  13.33%
      !              *                !   9   0.3  -  0.2       0   0.00%
   5-                *  *             -    A   0.4  -  0.3       0   0.00%
      !              *  *             !   B   0.5  -  0.4       0   0.00%
      !              *  *             !   C   ++   -  0.5       0   0.00%
      !           *  *  *             !     - - - - - - - - - - - - - - - -
      !           *  *  *             !          TOTAL         15 100.00%
     - - - - - - - - - - - - - - - - - - - -
         1  2  3  4  5  6  7  8  9  A  B  C     EACH "*" REPRESENTS 1 RESIDUALS
```

GOODNESS OF FIT SUMMARY

INDEPENDENCE MODEL CHI-SQUARE =   978.875 ON   6 DEGREES OF FREEDOM
INDEPENDENCE AIC =   966.87451   INDEPENDENCE CAIC =   942.38608
         MODEL AIC =   1.03236   MODEL CAIC =   -23.45607

CHI-SQUARE =   13.032 BASED ON   6 DEGREES OF FREEDOM
PROBABILITY VALUE FOR THE CHI-SQUARE STATISTIC IS   0.04252
THE NORMAL THEORY RLS CHI-SQUARE FOR THIS ML SOLUTION IS   12.945.

BENTLER-BONETT NORMED   FIT INDEX=   0.987
BENTLER-BONETT NONNORMED FIT INDEX=   0.993
COMPARATIVE FIT INDEX   =   0.993

As discussed earlier, the chi-square value is equal to $T = 13.032$ with df = 6 (note that there is a slight difference in the chi-square value from the

one provided by LISREL, due to rounding errors and nonidentical imple-
mentations of numerical minimization algorithms for the fit function).

ITERATIVE SUMMARY

ITERATION	PARAMETER ABS CHANGE	ALPHA	FUNCTION
1	1293.185180	1.00000	1903.47144
2	1069.007570	0.50000	15.59979
3	506.857086	1.00000	9.38928
4	118.013252	1.00000	7.51161
5	6.900131	1.00000	5.85625
6	57.521423	0.50000	4.72170
7	89.982933	0.50000	3.49870
8	98.234856	1.00000	2.13066
9	76.633858	1.00000	0.61933
10	22.431816	1.00000	0.08780
11	1.521491	1.00000	0.08147
12	0.070871	1.00000	0.08145
13	0.005703	1.00000	0.08145
14	0.000591	1.00000	0.08145

After an initially fairly large fit-function value, a subsequent clean convergence to the final solution is observed, which is what eventually mat-
ters for this example.

MEASUREMENT EQUATIONS WITH STANDARD ERRORS AND TEST STATISTICS

```
THU_IND1=V1    =     1.000 F1    + 1.000 E1


THU_IND2=V2    =     1.391*F1    + 1.000 E2
                      .024
                     58.355

THU_IND3=V3    =     1.434*F1    + 1.000 E3
                      .026
                     55.712

THU_IND4=V4    =     1.391*F1    + 1.000 E4
                      .024
                     58.355
```

CONSTRUCT EQUATIONS WITH STANDARD ERRORS AND TEST STATISTICS

```
TIME_FAC=F1    =    38.138*V999   + 1.000 D1
                     1.250
                    30.507
```

The latent mean is estimated at 38.138. As discussed earlier for the EQS
input file, this is the estimated slope of the regression of the latent factor

on the dummy variable V999 (i.e., the value of the only element of the AL-PHA matrix in the LISREL output). Next comes the final model output part that includes information about the variances of the independent variables and the standardized solution.

```
VARIANCES OF INDEPENDENT VARIABLES
----------------------------------

                  E                                    D
                 ---                                  ---
E1 -THU_IND1         71.996*I  D1 -TIME_FAC           185.691*I
                      8.440 I                          21.974 I
                      8.530 I                           8.450 I
                            I                                 I

E2 -THU_IND2         14.390*I                                I
                      3.078 I                                I
                      4.674 I                                I
                            I                                I

E3 -THU_IND3         25.384*I                                I
                      4.040 I                                I
                      6.284 I                                I
                            I                                I

E4 -THU_IND4         27.511*I                                I
                      4.118 I                                I
                      6.680 I                                I
                            I                                I


STANDARDIZED SOLUTION:

THU_IND1=V1  =     .849 F1        +      .529 E1
THU_IND2=V2  =     .981*F1        +      .196 E2
THU_IND3=V3  =     .968*F1        +      .250 E3
THU_IND4=V4  =     .964*F1        +      .267 E4
TIME_FAC=F1  =     .000*V999      +     1.000 D1
```

LEVEL AND SHAPE MODEL

The one-factor LCA model presented in the previous sections is a rather restricted model because only one factor is postulated to account for change over time. The level and shape (LS) model was first described in the late 1980s and popularized by McArdle and colleagues as a very useful two-factor LCA model in longitudinal research (McArdle, 1988; McArdle & Anderson, 1990). The LS model was specifically developed to study two important aspects of the process of latent change: (a) initial ability status at the beginning of a study (i.e., the Level factor), and (b) the change (increase or decrease) in ability status at the end of a study (i.e., the Shape factor). An example path diagram of a LS model with four successive measurement occasions (Y's) is presented in Fig. 13 using LISREL notation.

As can be seen by examining the model presented in Fig. 13, each observed variable Y loads on two factors, the Level and Shape factors (i.e., η_1 = Initial status factor, and η_2 = Change factor). Fitting the LS model to data and interpreting the Level and Shape factors are accomplished by fixing the majority of the factor loadings to a value of 1, as indicated in Fig. 13 by the 1s attached to most of the paths. Fixing the loadings of the four assessment occasions on the level factor to 1 ensures that it is interpreted as an initial ability status factor (i.e., as a baseline level of the developmental process under investigation). Fixing the loading of the last assessment occasion on the Shape factor to 1 as well and that of the first assessment occasion on it to 0, ensures that this factor is interpreted as an overall ability change factor (i.e., shape of the change process). Freeing the loadings of the second and third assessment occasions on the Shape factor implies that they denote the part of overall ability change that occurs between the first and each of these later two measurement occasions (McArdle & Anderson, 1990). Finally, the Level and Shape factors are assumed to be correlated, although this need not always be the case.

Using this approach, the LS model achieves something that the simple one-factor model presented earlier cannot—the modeling of individual differences in change over the repeated assessment occasions, which are reflected in its Shape factor. The reason for this limitation in the

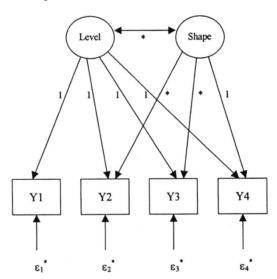

FIG. 13.   Example two-factor LCA model.

one-factor model is that it cannot effectively separate the initial status from change in the studied construct. This limitation becomes particularly important when one is interested in studying correlates and predictors of growth or decline. In such instances, as shown in a later section, the LS model amplifies its advantages further.

Now to demonstrate an analysis based on the LS model, consider the data from the cognitive intervention study used for the one-factor model. Recall that four repeated assessments on $N = 161$ older adults were carried out using Thurstone's Standard Induction test (a) before the administration of the training program (Y_1), and then (b) 1 week after (Y_2), (c) 1 month after (Y_3), and (d) 6 months after (Y_4) completion of training. To determine the parameters of the model presented in Fig. 13, one must again follow the six rules outlined in chapter 1. Rules 1 and 2 imply that all error variances and latent variances and covariances are parameters. Rule 3 suggests that all factor loadings are parameters (except of course those that are already specified as fixed). Rule 4 is not applicable in this model because there are no explanatory relationships among the latent variables. Rule 6 is also not applicable in this model because, by fixing most of the factor loadings to 1, the metric of each latent variable is already set.

LISREL AND EQS INPUT FILES

The input files for both the LISREL and EQS programs are constructed by a simple modification of the input files for the one-factor LCA model presented earlier. Only two changes are needed, adding information about the presence of a second factor in the model and then fixing the appropriate factor loadings. Making these changes provides the following LISREL and EQS input files. Note that in order to keep the error impacts on test performance equal across time and at the same time arrive at a more parsimonious model, the error variances are set to EQual one another in LISREL, and that /CONSTRAINTS command is introduced in EQS.

LISREL Input File

```
LEVEL AND SHAPE MODEL
DA NI=4 NO=161
CM
307.46
296.52   377.21
295.02   365.10   392.47
```

```
291.02   355.88   358.25   376.84
ME
37.48   53.30   54.82   52.63
LA
THU_IND1 THU_IND2 THU_IND3 THU_IND4
MO NY=4 NE=2 AL=FR PS=SY,FR
LE
LEVEL SHAPE
VA 1 LY(1, 1) LY(2, 1) LY(3, 1) LY(4, 1)
VA 1 LY(4, 2)
FR LY(2, 2) LY(3, 2)
EQ TE(1)-TE(4)
OU
```

EQS Input File

```
/TITLE
LEVEL AND SHAPE MODEL;
/SPECIFICATIONS
VARIABLES=4; CASES=161;
/LABELS
LA
V1=THU_IND1; V2=THU_IND2; V3=THU_IND3; V4=THU_IND4;
F1=LEVEL; F2=SHAPE;
/EQUATIONS
V1=F1+E1;
V2=F1+*F2+E2;
V3=F1+*F2+E3;
V4=F1+F2+E4;
F1=*V999+D1;
F2=*V999+D2;
/VARIANCES
D1 TO D2=*; E1 TO E4=*;
/COVARIANCES
D1,D2=*;
/CONSTRAINTS
(E1,E1)=(E2,E2)=(E3,E3)=(E4,E4);
/MATRIX
307.46
296.52   377.21
295.02   365.10   392.47
291.02   355.88   358.25   376.84
```

```
/MEANS
37.48   53.30   54.82   52.63
/END;
```

MODELING RESULTS

LISREL Program Results

The output produced by the LISREL input file in the previous section is presented next. (As before, comments are inserted at appropriate places to clarify portions of the output and some sections have been omitted.)

```
PARAMETER SPECIFICATIONS

          LAMBDA-Y

               LEVEL      SHAPE
               -----      -----
THU_IND1         0          0
THU_IND2         0          1
THU_IND3         0          2
THU_IND4         0          0

          PSI

               LEVEL      SHAPE
               -----      -----
LEVEL            3
SHAPE            4          5

          THETA-EPS

          THU_IND1   THU_IND2   THU_IND3   THU_IND4
          --------   --------   --------   --------
             6          6          6          6

          ALPHA

               LEVEL      SHAPE
               -----      -----
                 7          8
```

Observe from this section that the number and location of the parameters have been correctly communicated to the LISREL program.

```
              GOODNESS OF FIT STATISTICS

   CHI-SQUARE WITH 6 DEGREES OF FREEDOM = 6.79 (P = 0.34)
      ESTIMATED NON-CENTRALITY PARAMETER (NCP) = 0.79
   90 PERCENT CONFIDENCE INTERVAL FOR NCP = (0.0 ; 11.53)
```

```
                 MINIMUM FIT FUNCTION VALUE = 0.042
         POPULATION DISCREPANCY FUNCTION VALUE (F0) = 0.0049
       90 PERCENT CONFIDENCE INTERVAL FOR F0 = (0.0 ; 0.072)
      ROOT MEAN SQUARE ERROR OF APPROXIMATION (RMSEA) = 0.029
     90 PERCENT CONFIDENCE INTERVAL FOR RMSEA = (0.0 ; 0.11)
         P-VALUE FOR TEST OF CLOSE FIT (RMSEA < 0.05) = 0.57


           EXPECTED CROSS-VALIDATION INDEX (ECVI) = 0.14
      90 PERCENT CONFIDENCE INTERVAL FOR ECVI = (0.11 ; 0.18)
                 ECVI FOR SATURATED MODEL = 0.13
                ECVI FOR INDEPENDENCE MODEL = 6.17


CHI-SQUARE FOR INDEPENDENCE MODEL WITH 6 DEGREES OF FREEDOM = 978.84
                    INDEPENDENCE AIC = 986.84
                        MODEL AIC = 22.79
                      SATURATED AIC = 20.00
                   INDEPENDENCE CAIC = 1003.17
                       MODEL CAIC = 55.44
                     SATURATED CAIC = 60.81


             ROOT MEAN SQUARE RESIDUAL (RMR) = 1.80
                   STANDARDIZED RMR = 0.0050
               GOODNESS OF FIT INDEX (GFI) = 0.98
          ADJUSTED GOODNESS OF FIT INDEX (AGFI) = 0.97
         PARSIMONY GOODNESS OF FIT INDEX (PGFI) = 0.59


                 NORMED FIT INDEX (NFI) = 0.99
               NON-NORMED FIT INDEX (NNFI) = 1.00
           PARSIMONY NORMED FIT INDEX (PNFI) = 0.99
              COMPARATIVE FIT INDEX (CFI) = 1.00
              INCREMENTAL FIT INDEX (IFI) = 1.00
               RELATIVE FIT INDEX (RFI) = 0.99


                   CRITICAL N (CN) = 397.43
```

All of the goodness-of-fit indices support an acceptable model fit (and although they are not presented, the model residuals support the same conclusion). Note in particular the relatively low chi-square value (observe its degrees of freedom and p value), as well as the fairly low RMSEA. In addition, note that the left endpoint of its confidence interval as well as those of the noncentrality parameter and population fit function minimum are 0.

It is important to note that the LS model has a much better fit than the one-factor LCA model that is fitted to this data in the preceding section. The reason is that the LS model uses two factors to account for change and, as a consequence, uses two factor means that specifically account for the growth in the variable means beyond factor loadings explaining the structure of their covariance matrix.

```
LISREL ESTIMATES (MAXIMUM LIKELIHOOD)


        LAMBDA-Y


                  LEVEL        SHAPE
                  -----        -----
THU_IND1           1.00         - -

THU_IND2           1.00         1.03
                               (.03)
                               31.41

THU_IND3           1.00         1.14
                               (.03)
                               32.70

THU_IND4           1.00         1.00
```

Using again the CI approach presented in the discussion of the one-factor model, note that there is some indication of a marked ability change from the first assessment to the second assessment occasion (subtract twice the standard error from and add twice the standard error to parameter estimate to obtain the pertinent CI). Indeed, at the second assessment occasion, the CI of the loading on the Shape factor is (.97; 1.09), whereas the loading on the first assessment occasion is the constant 0 (and hence completely to the left of this CI). The same result holds for the remaining assessment occasions—ability at any of them is markedly higher than that at the first assessment occasion. In fact, ability growth seems highest at the third assessment occasion—the CI of its loading on the Shape factor is (1.08; 1.20) and only marginally overlaps with the CI of its loading on the Shape factor at the second assessment occasion. Also, because the Shape factor loading for the fourth assessment occasion is set equal to 1 (and thus completely to the left of the CI), one can conclude that a considerable decline in test performance occurred from the third to the fourth assessment occasion. Interestingly, this finding was also observed with the single-factor model in Fig. 12, perhaps reflecting a wearing off of the cognitive intervention effect.

```
COVARIANCE MATRIX OF ETA


                  LEVEL        SHAPE
                  -----        -----
LEVEL            285.00
SHAPE              8.74        50.39
```

```
            PSI

            LEVEL        SHAPE
            -----        -----
LEVEL       285.00
            (34.38)
              8.29

SHAPE         8.74       50.39
            (12.28)      (9.06)
               .71        5.56
```

Note that the covariance between the Level and Shape factors is nonsignificant. Such a finding is possible in empirical research. Substantively, it indicates that the starting position on the Induction construct is not necessarily always predictive of the amount of growth along this dimension. Of course, one should interpret this finding with some caution, especially because the sample size used in the study is only 161 and cannot really be considered large (and is used here only for illustrative purposes).

```
            THETA-EPS


            THU_IND1    THU_IND2    THU_IND3    THU_IND4
            --------    --------    --------    --------
              22.29       22.29       22.29       22.29
             (1.76)      (1.76)      (1.76)      (1.76)
              12.65       12.65       12.65       12.65


            SQUARED MULTIPLE CORRELATIONS FOR Y - VARIABLES

            THU_IND1    THU_IND2    THU_IND3    THU_IND4
            --------    --------    --------    --------
                .93         .94         .94         .94

            ALPHA

            LEVEL        SHAPE
            -----        -----
            37.48        15.24
            (1.39)       (.76)
            27.04        20.03
```

The ALPHA estimates provided here represent the mean initial ability status and mean overall ability change across the four assessment occasions of the study. The results indicate that the average subject performance is substantially higher at the end of the study, and hence the cognitive training had a lasting effect.

EQS Program Results

The EQS input file described in the previous section produces the following results for the LS model examined.

```
MAXIMUM LIKELIHOOD SOLUTION (NORMAL DISTRIBUTION THEORY)

PARAMETER ESTIMATES APPEAR IN ORDER,
NO SPECIAL PROBLEMS WERE ENCOUNTERED DURING OPTIMIZATION.

ALL EQUALITY CONSTRAINTS WERE CORRECTLY IMPOSED
```

As usual, this portion of the output is inspected in order to ensure that there are no problems with the implementation of the model, it is technically sound, and its coding into the EQS notation is correct.

```
RESIDUAL COVARIANCE/MEAN MATRIX   (S-SIGMA):
```

		THU_IND1 V 1	THU_IND2 V 2	THU_IND3 V 3	THU_IND4 V 4	V999 V999
THU_IND1 V	1	0.172				
THU_IND2 V	2	2.487	-1.968			
THU_IND3 V	3	0.078	1.879	0.091		
THU_IND4 V	4	-2.719	1.031	-2.750	1.693	
V999	V999	0.004	0.073	0.008	-0.085	0.000

```
            AVERAGE ABSOLUTE COVARIANCE RESIDUALS          =       1.0026
   AVERAGE OFF-DIAGONAL ABSOLUTE COVARIANCE RESIDUALS      =       1.1114
```

```
STANDARDIZED RESIDUAL MATRIX:
```

		THU_IND1 V 1	THU_IND2 V 2	THU_IND3 V 3	THU_IND4 V 4	V999 V999
THU_IND1 V	1	0.001				
THU_IND2 V	2	0.007	-0.005			
THU_IND3 V	3	0.000	0.005	0.000		
THU_IND4 V	4	-0.008	0.003	-0.007	0.004	
V999	V999	0.000	0.004	0.000	-0.004	0.000

```
            AVERAGE ABSOLUTE STANDARDIZED RESIDUALS        =       0.0033
   AVERAGE OFF-DIAGONAL ABSOLUTE STANDARDIZED RESIDUALS    =       0.0039
```

```
LARGEST STANDARDIZED RESIDUALS:

   V  4,V  1      V  2,V  1      V  4,V  3      V  2,V  2      V  3,V  2
     -0.008          0.007         -0.007         -0.005          0.005

   V  4,V  4     V999,V  4      V999,V  2      V  4,V  2      V  1,V  1
      0.004         -0.004          0.004          0.003          0.001

   V999,V  3     V  3,V  3      V  3,V  1      V999,V  1      V999,V999
      0.000          0.000          0.000          0.000          0.000
```

DISTRIBUTION OF STANDARDIZED RESIDUALS

```
     - - - - - - - - - - - - - - - - - - - - - - - - - - -
     !                              !
  20 -                              -
     !                              !
     !                              !
     !                              !
     !                              !         RANGE        FREQ  PERCENT
  15 -                              -
     !                              !   1   -0.5  -  --       0    0.00%
     !                              !   2   -0.4  -  -0.5     0    0.00%
     !                              !   3   -0.3  -  -0.4     0    0.00%
     !                              !   4   -0.2  -  -0.3     0    0.00%
  10 -               *              -   5   -0.1  -  -0.2     0    0.00%
     !               *              !   6    0.0  -  -0.1     5   33.33%
     !               *              !   7    0.1  -   0.0    10   66.67%
     !               *              !   8    0.2  -   0.1     0    0.00%
     !               *              !   9    0.3  -   0.2     0    0.00%
   5 -           *   *              -   A    0.4  -   0.3     0    0.00%
     !           *   *              !   B    0.5  -   0.4     0    0.00%
     !           *   *              !   C    ++   -   0.5     0    0.00%
     !           *   *              !       - - - - - - - - - - - - - - - - - - - - - - -
     !           *   *              !          TOTAL         15  100.00%
     - - - - - - - - - - - - - - - - - - - - - - - - - -
       1  2  3  4  5  6  7  8  9  A  B  C   EACH "*" REPRESENTS  1 RESIDUALS
```

Based on this (and as concluded for the LISREL output), none of the standardized residuals is large (i.e., the LS model accounts quite well for all variable variances and covariances).

```
GOODNESS OF FIT SUMMARY

INDEPENDENCE MODEL CHI-SQUARE =         978.875 ON      6 DEGREES OF FREEDOM

INDEPENDENCE AIC =    966.87451   INDEPENDENCE CAIC =    942.38608
        MODEL AIC =     -5.20723          MODEL CAIC =    -29.69566

CHI-SQUARE =        6.793 BASED ON     6 DEGREES OF FREEDOM
PROBABILITY VALUE FOR THE CHI-SQUARE STATISTIC IS      0.34044
HE NORMAL THEORY RLS CHI-SQUARE FOR THIS ML SOLUTION IS             6.630.

BENTLER-BONETT NORMED    FIT INDEX=      0.993
BENTLER-BONETT NONNORMED FIT INDEX=      0.999
COMPARATIVE FIT INDEX             =      0.999
```

Once again, all goodness-of-fit indices support a well-fitting model whose normed, nonnormed, and comparative fit indices are all very close to 1.

```
                        ITERATIVE SUMMARY
                   PARAMETER
ITERATION          ABS CHANGE              ALPHA            FUNCTION
    1              1262.579590            1.00000            7.83688
    2               201.485840            1.00000            3.73936
    3                14.295440            1.00000            2.21572
    4                 8.654326            1.00000            0.67292
    5                13.698804            1.00000            0.05055
    6                 1.476818            1.00000            0.04246
    7                 0.005894            1.00000            0.04245
    8                 0.001118            1.00000            0.04246
    9                 0.000006            1.00000            0.04245
```

Observe that a direct convergence to the final solution was reached within fewer than 10 iterations.

```
MEASUREMENT EQUATIONS WITH STANDARD ERRORS AND TEST STATISTICS

THU_IND1=V1    =    1.000 F1      + 1.000 E1


THU_IND2=V2    =    1.000 F1      + 1.034*F2       + 1.000 E2
                                    .033
                                  31.413

THU_IND3=V3    =    1.000 F1      + 1.138*F2       + 1.000 E3
                                    .035
                                  32.703

THU_IND4=V4    =    1.000 F1      + 1.000 F2       + 1.000 E4


   CONSTRUCT EQUATIONS WITH STANDARD ERRORS AND TEST STATISTICS

LEVEL =F1      =    37.476*V999   + 1.000 D1
                    1.386
                   27.044

SHAPE =F2      =    15.238*V999   + 1.000 D2
                     .761
                   20.028


VARIANCES OF INDEPENDENT VARIABLES
- - - - - - - - - - - - - - - - - - - - - - - - - -
                         E                          D
                         - - -                      - - -
E1 -THU_IND1             22.285*I D1 -LEVEL          285.003*I
                          1.762 I                    34.378 I
                         12.649 I                     8.290 I
                                I                            I
E2 -THU_IND2             22.285*I D2 -SHAPE           50.386*I
                          1.762 I                     9.062 I
                         12.649 I                     5.560 I
                                I                            I
```

```
E3  -THU_IND3              22.285*I                                I
                            1.762 I                                I
                           12.649 I                                I
                                  I                                I
E4  -THU_IND4              22.285*I                                I
                            1.762 I                                I
                           12.649 I                                I
                                  I                                I
```

COVARIANCES AMONG INDEPENDENT VARIABLES
- -

```
              E                           D
             - - -                       - - -
                        I D2 -SHAPE              8.736*I
                        I D1 -LEVEL             12.278 I
                        I                         .712 I
                        I                              I
```

CORRELATIONS AMONG INDEPENDENT VARIABLES
- -

```
              E                           D
             - - -                       - - -
                        I D2 -SHAPE               .073*I
                        I D1 -LEVEL                    I
                        I                              I
```

Note again the nonsignificant covariance between the Level and Shape factors (and recall the earlier discussion of this result). A nice feature of the EQS program is that it also presents this estimated value as a correlation coefficient between independent variables (although rather weak at .073). (It is important to recall that the asterisk attached to this correlation estimate denotes only that the pertinent covariance has been estimated by the program as a model parameter, not that it is significant.)

STUDYING CORRELATES AND PREDICTORS
OF LATENT CHANGE

The LS model is also very useful for examining correlates and predictors of change. Such concerns arise frequently in practice when a researcher is interested in determining the specific behavioral or social characteristics of individuals that are related to the observed patterns of change. For example, in the cognitive intervention study of older adults presented earlier, there may be an interest in finding out which kinds of individuals exhibit the most salient improvement or least pronounced decline along the dimensions studied. The special property of the LS model that makes it such a useful means for addressing these types of queries is the fact that it parameterizes overall ability change in one of its latent variables, viz. the Shape factor.

To permit the study of correlates and predictors of change, the LS model is extended to include covariates and relate them to the Level and Shape factors. Specifically, by focusing on the correlations of the Shape factor with putative predictors, one obtains estimates of the degrees of interrelationship between the latter and overall latent change. For example, if these correlations are notable and positive, individuals with high values on the correlates tend to be among those who improve most, and individuals with low values tend to improve the least. Conversely, if the correlations are negative, individuals with high correlate values tend to be among those who improve least, and individuals with low correlate values tend to improve the most. Alternatively, when studying correlates of decline, if these correlations are marked and positive, then individuals with high correlate values tend to be among those who decline least, and individuals with low covariate values tend to decline the most. If the correlations are negative, individuals with high correlate values tend to be among those who decline the most, and individuals with low correlate values tend to decline the least.

This extension to the LS model is displayed graphically in Fig. 14 for the case of a latent covariate (η_3) with two indicators (Y_5 and Y_6) using the cognitive intervention study on older adults presented earlier, which is examined further in the next section. It is important to note that the only difference between the LS model presented in Fig. 14 and the LS model in Fig. 13 is the added latent covariate. It must be emphasized that for the purposes of studying predictors of change one can include any number of covariates in an appropriately extended LS model. In fact, the covariates may be (a) manifest or latent, (b) perfectly measured or fallible, or (c) once or repeatedly assessed. In the case of (c), one may also consider modeling the repeated assessments of the covariates themselves in terms of a pertinent LS model (e.g., Raykov, 1995).

Before proceeding to the analyses of the cognitive intervention data (discussed in the sections on the one-factor LCA model) based on the LS model in Fig. 14, a useful generalization of the modeling approach, for studying multiple groups called a multisample analysis, is introduced.

Multisample Analysis

Many studies, especially in the behavioral and social sciences, examine differences or similarities between two or more groups in the structure of a phenomenon under investigation. For example, groups may differ from one another in terms of age, educational level, nationality, ethnicity, religious affiliation, or political affiliation. When the same phenomenon is studied in all of them, it is important to have a methodological means that allows researchers to compare the groups along special dimensions of interest as well as to pinpoint their similarities or lack thereof. SEM offers a

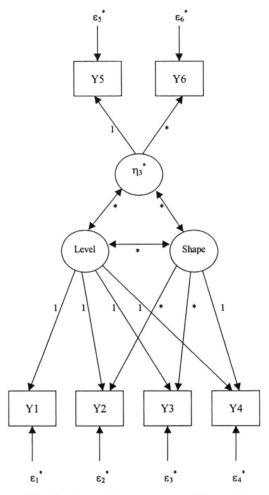

FIG. 14.   Example latent covariate LCA model.

method for conducting these types of comparisons using a modification of its general model-fitting approach. This extension accounts for the fact that group comparisons necessitate the simultaneous estimation of models in all the samples involved. To this end, the model(s) of interest are postulated within each of the groups and then their simultaneous estimation is carried out. This is accomplished by minimizing a compound fit function that results by adding the fit functions across the groups, thereby weighting them proportionately to the sizes of the available samples. This permits the simultaneous estimation of all parameters of the models in all groups. At the minimum of that fit function, a test of the overall model is possible, just as in the case of a single population (see chap. 1). It is also possible to impose restrictions on model parameters within as well as across groups, and to estimate and test the overall model subject to these constraints. This is achieved in a way that is very similar to fitting a model to a single group, and accounts for the fact that one is dealing with an extended simultaneous model across all groups in which the (cross-group) restricted parameters satisfy the imposed constraint(s).

Correlates of Change in the Cognitive Intervention Study

In this section the multisample approach is demonstrated for the extended LS model for studying the correlates of change presented in Fig. 14. It is applied to data based partly on the Baltes et al. (1986) study, which was originally conceptualized and conducted as a two-group investigation. In the previous sections, only the data from the experimental group ($N = 161$) was analyzed. There was also a no-contact control group with $N = 87$ older adults, who did not receive any cognitive training. All members of the control group were administered the same test battery at the same assessment points as the experimental group. Now a two-group comparison can be conducted to determine whether the Figural relations construct can be considered to be a covariate of performance improvement. Two measures are used as indicators of the Figural relations construct, a figural relations test, and a culture-fair test. By conducting the multisample analysis, one is able to do the following: (a) ascertain if there is improvement in performance in the control and experimental groups, (b) examine group similarities and differences in the pattern of change across the 6 months of the study, (c) see if pretest Figural relations is predictive of subsequent improvement in Induction test performance, and (d) explore whether the two groups differ with respect to point c.

Input File, Multisample Analysis

The LISREL program must be advised that a multisample analysis is needed. In order to convey this information, a number of new commands lines are introduced in the following input file.

```
STUDYING COVARIATES AND PREDICTORS OF CHANGE * EXP. GR.
DA NI=6 NO=161 NG=2
CM
307.46
296.52   377.21
295.02   365.10   392.47
291.02   355.88   358.25   376.84
203.47   229.25   221.49   229.92   260.76
169.40   180.76   185.95   181.70   155.22   159.38
```

```
ME
37.48    53.30    54.82    52.63    49.48    48.39
LA
THU_IND1 THU_IND2 THU_IND3 THU_IND4 FR11 FR12
MO NY=6 NE=3 AL=FR PS=SY,FR
LE
LEVEL SHAPE FIG_REL1
VA 1 LY(1, 1) LY(2, 1) LY(3, 1) LY(4, 1)
VA 1 LY(4, 2)
FR LY(2, 2) LY(3, 2)
VA 1 LY(5, 3)
FR LY(6, 3)
EQ TE(1)-TE(4)
ST .1 ALL
ST 100 PS(1, 1) PS(2, 2) PS(3, 3) TE(1)-TE(6)
OU NS
* CONTROL GROUP
DA NI=35 NO=87
CM
319.22
326.42    410.25
318.80    378.98    409.07
333.06    388.68    390.59    432.70
171.01    209.41    215.96    213.76    262.79
149.82    192.75    193.69    200.82    148.86    161.98
ME
38.62    44.96    47.88    48.01    51.97    49.81
LA
THU_IND1 THU_IND2 THU_IND3 THU_IND4 FR11 FR12
MO LY=PS PS=PS TE=PS AL=PS
LE
LEVEL SHAPE FIG_REL1
OU
```

The LISREL input file has two main parts that pertain to the model specification for each of the two groups under consideration. Note that in the title line (which will recur as a page title throughout the output for the first group) there is an indication that the first portion of the input contains the information for the experimental group. The keyword NG=2 advises the program of the fact that a two-group analysis is to be performed (NG for Number of Groups). Also note that the number of analyzed variables is now six because the earlier LS model has been extended to include the two indicators of the covariate factor, as presented in Fig. 14. The remaining portion of the first part of the input file is largely the same as that for the single-group LS model presented in the previous section; this is because the portion defines exactly the same model for the experimental group. The only difference is that here there are three latent variables—in addition to the Level and Shape factors, Figural relations is also included in the model.

The three latent means are next declared to be model parameters by the keyword AL=FR. Then the latent covariance matrix is defined as having all its elements as parameters by the keyword PS=SY,FR (i.e., as being symmetric and consisting of free parameters only). The remaining three matrices (LY, BE, and TE) are not mentioned in this line because they should be at their defaults. (At the default settings, LY is full and fixed; BE is a zero matrix, i.e., there are no latent explanatory relationships; and TE is diagonal and free, i.e., having model parameters along its diagonal that are all error variances and parameters.) The free factor loadings are then declared: one for a Figural relations indicator, using the keyword FR LY(5, 3); and two pertaining to the repeated assessments with the Induction test. Next the scale of the Figural relations construct is set by fixing its other factor loading to 1.

After imposing the constraint on the corresponding four error variances, one provides start values for all the model parameters. This activity is sometimes needed in practice because structural modeling programs are not always able to come up with their own good start values. Thus they need an initial jump start to find the road toward the final solution. To achieve this, first start all parameters at a value of 0.1 and then, with the next line, override this choice only for the variances by starting them at a value of 100. This choice of large start values for all variance parameters (pertaining to both latent and observed variables) and small values for all the other model parameters (including factor loadings and independent variable covariances) is in general a good way of providing initial values for model parameters. It actually contributes to the stability of the numerical subroutine during the fit-function minimization process. The last line pertaining to the experimental group signals the end of the input by the keyword OU, and asks the program to use the specifically supplied start values instead of computing its own, as it would otherwise. This activity is achieved by the added keyword NS (for No Start values), which indicates that the program should not use its computed start values, but rather those provided by the user.

The next input line provides a title for the output portions pertaining to the results for the second group (i.e., the control group). This title line will also appear at the beginning of any output page that presents results of analyses for the control group. Thus, it is advisable to always have a title line to label each group in a multisample analysis. The next data line advises the program of the number of variables underlying the following covariance matrix and means row, as well as sample size in the control group. The model line then notifies the program that in this group a model is being defined in which all matrices have the same pattern and parameters as those of the preceding group (i.e., the experimental group). This is achieved by stating each matrix name followed by the keyword PS (for same Pattern and same Start values; unmentioned matrices are assumed to be invariant across groups). The effect of this is that exactly the same model is defined in the control group as in the experimental one, and its parameters are assigned the same initial values as in that group. The last line OU signals the end of the input file for the current group, and hence for the whole model because it is being fitted to two groups.

Modeling Results, Multisample Analysis

The LISREL input file generates the following output (again, omitting irrelevant or redundant sections and inserting comments where appropriate).

```
EXPERIMENTAL GROUP

PARAMETER SPECIFICATIONS

        LAMBDA-Y

                  LEVEL       SHAPE     FIG_REL1
                  -----       -----     --------
THU_IND1            0           0           0
THU_IND2            0           1           0
THU_IND3            0           2           0
THU_IND4            0           0           0
    FR11            0           0           0
    FR12            0           0           3

        PSI

                  LEVEL       SHAPE     FIG_REL1
                  -----       -----     --------
   LEVEL            4
   SHAPE            5           6
FIG_REL1            7           8           9

        THETA-EPS

              THU_IND1    THU_IND2    THU_IND3    THU_IND4      FR11        FR12
              --------    --------    --------    --------      ----        ----
                 10          10          10          10          11          12

        ALPHA

                  LEVEL       SHAPE     FIG_REL1
                  -----       -----     --------
                   13          14          15
```

```
* CONTROL GROUP

PARAMETER SPECIFICATIONS

        LAMBDA-Y

                LEVEL       SHAPE    FIG_REL1
                -----       -----    --------
THU_IND1            0           0           0
THU_IND2            0          16           0
THU_IND3            0          17           0
THU_IND4            0           0           0
    FR11            0           0           0
    FR12            0           0          18

        PSI

                LEVEL       SHAPE    FIG_REL1
                -----       -----    --------
   LEVEL           19
   SHAPE           20          21
FIG_REL1           22          23          24

        THETA-EPS

            THU_IND1    THU_IND2    THU_IND3    THU_IND4        FR11        FR12
            --------    --------    --------    --------        ----        ----
                  25          25          25          25          26          27

        ALPHA

                LEVEL       SHAPE    FIG_REL1
                -----       -----    --------
                   28          29          30
```

First, make sure that the numbering and positions of the model parameters are correct for both groups. Note that the numbering of model parameters begins in the experimental groups and continues in the control group, rather than starting anew. This is due to the fact that LISREL has been programmed to perform a simultaneous analysis rather than two separate analyses. This feature will become especially important in the next fitted model, when cross-group parameter constraints are introduced.

Before beginning the parameter estimate interpretation, inspect the goodness-of-fit section of the output.

```
                    GOODNESS OF FIT STATISTICS

        CHI-SQUARE WITH 24 DEGREES OF FREEDOM = 27.33 (P = 0.29)

                CONTRIBUTION TO CHI-SQUARE = 4.91
            PERCENTAGE CONTRIBUTION TO CHI-SQUARE = 17.98
            ESTIMATED NON-CENTRALITY PARAMETER (NCP) = 3.33
        90 PERCENT CONFIDENCE INTERVAL FOR NCP = (0.0 ; 20.46)
```

```
MINIMUM FIT FUNCTION VALUE = 0.11
POPULATION DISCREPANCY FUNCTION VALUE (F0) = 0.014
90 PERCENT CONFIDENCE INTERVAL FOR F0 = (0.0 ; 0.083)
ROOT MEAN SQUARE ERROR OF APPROXIMATION (RMSEA) = 0.024
90 PERCENT CONFIDENCE INTERVAL FOR RMSEA = (0.0 ; 0.059)
P-VALUE FOR TEST OF CLOSE FIT (RMSEA < 0.05) = 0.87


EXPECTED CROSS-VALIDATION INDEX (ECVI) = 0.36
90 PERCENT CONFIDENCE INTERVAL FOR ECVI = (0.21 ; 0.29)
ECVI FOR SATURATED MODEL = 0.17
ECVI FOR INDEPENDENCE MODEL = 8.07


CHI-SQUARE FOR INDEPENDENCE MODEL WITH 30 DEGREES OF FREEDOM = 1972.71
INDEPENDENCE AIC = 1996.71
MODEL AIC = 87.33
SATURATED AIC = 84.00
INDEPENDENCE CAIC = 2050.88
MODEL CAIC = 222.73
SATURATED CAIC = 273.56


ROOT MEAN SQUARE RESIDUAL (RMR) = 8.43
STANDARDIZED RMR = 0.024
GOODNESS OF FIT INDEX (GFI) = 0.98
PARSIMONY GOODNESS OF FIT INDEX (PGFI) = 1.12


NORMED FIT INDEX (NFI) = 0.99
NON-NORMED FIT INDEX (NNFI) = 1.00
PARSIMONY NORMED FIT INDEX (PNFI) = 0.79
COMPARATIVE FIT INDEX (CFI) = 1.00
INCREMENTAL FIT INDEX (IFI) = 1.00
RELATIVE FIT INDEX (RFI) = 0.98
CRITICAL N (CN) = 387.84
```

All indices point to an acceptable fit of the overall two-group model. Note the low magnitude of the chi-square value compared to the model degrees of freedom (observe its p value), as well as the RMSEA index, which is well below the proposed threshold level of .05. Furthermore, the left endpoints of the confidence intervals for RMSEA, the noncentrality parameter, and the population minimal fit-function value are all at their best value of 0. It is important to note that the degrees of freedom are twice those for the extended LS model when it is fitted to only a single group. The same holds true for the model parameters—their number is twice that of the extended LS model when fitted to only one group. This result will always be true whenever the same model is postulated in two distinct groups and no cross-group parameter constraints are imposed—both the degrees of freedom and number of parameters in the overall two-group model are exactly two times those if the model were fitted to either of the groups separately.

Given the satisfactory fit indices of the overall two-group model, one can now continue with parameter interpretation.

```
EXPERIMENTAL GROUP

LISREL ESTIMATES (MAXIMUM LIKELIHOOD)

        LAMBDA-Y

                  LEVEL        SHAPE      FIG_REL1
                  -----        -----      --------
THU_IND1          1.00          - -          - -

THU_IND2          1.00         1.03          - -
                              (0.03)
                              31.43

THU_IND3          1.00         1.14          - -
                              (0.03)
                              32.72

THU_IND4          1.00         1.00          - -

   FR11            - -          - -         1.00

   FR12            - -          - -         0.97
                                           (0.02)
                                           61.34
```

The first four lines of the factor loading matrix contain the same results as those in the preceding section in which the experimental group data was analyzed with the LS model, and hence warrant the same interpretations. The last two lines of the matrix pertain to the added covariate indicators and it is clear that they load quite similarly on the Figural relations factor—observe the standard error of the free factor loading and its directly obtained confidence interval (0.93; 1.01), which includes 1, the factor loading of the other latent covariate indicator.

```
COVARIANCE MATRIX OF ETA

                  LEVEL        SHAPE      FIG_REL1
                  -----        -----      --------
    LEVEL        285.00
    SHAPE          8.74        50.32
 FIG_REL1       182.51        15.44       150.63

        PSI

                  LEVEL        SHAPE      FIG_REL1
                  -----        -----      --------
    LEVEL        285.00
                (34.38)
                  8.29

    SHAPE          8.74        50.32
                (12.27)       (9.05)
                  0.71         5.56
```

```
FIG_REL1    182.51       15.44      150.63
           (23.32)      (9.17)     (19.92)
              7.83        1.68        7.56
```

From the last set of lines in the PSI matrix one can infer that there is a weak (linear) relationship between Figural relations as a putative correlate of change along the Induction dimension. The covariance between the shape factor and initial Figural relations falls short of statistical significance: estimate = 15.44, standard error = 9.17, and t value = 1.68. Although this result may be a consequence of the limited sample size (as mentioned before), it appears that even though the Figural relations construct represents a correlate of the initial status in inductive-reasoning ability, it has a limited capability for predicting the amount of improvement along the Induction dimension.

```
THETA-EPS
```

	THU_IND1	THU_IND2	THU_IND3	THU_IND4	FR11	FR12
	22.29	22.29	22.29	22.29	80.44	28.08
	(1.76)	(1.76)	(1.76)	(1.76)	(10.98)	(6.71)
	12.65	12.65	12.65	12.65	7.32	4.18

```
SQUARED MULTIPLE CORRELATIONS FOR Y - VARIABLES
```

	THU_IND1	THU_IND2	THU_IND3	THU_IND4	FR11	FR12
	.93	.94	.94	.94	.65	.83

```
ALPHA
```

	LEVEL	SHAPE	FIG_REL1
	37.48	15.23	49.78
	(1.39)	(.76)	(1.19)
	27.05	20.02	41.66

```
                 GOODNESS OF FIT STATISTICS

          CONTRIBUTION TO CHI-SQUARE = 22.42
     PERCENTAGE CONTRIBUTION TO CHI-SQUARE = 82.02
```

The goodness-of-fit statistics represent the chi-square contribution by the experimental group to the overall fit of the model. Although it is rather difficult to judge this contribution quantity in a simultaneous modeling analysis, one can always relate it to the degrees of freedom for a model involving no group constraints. In this case, the degrees of freedom are 12 (half these of the overall model) and so this chi-square value does not seem excessively large. Thus, informally, in the experimental group there is no indication of serious model misfit.

CONTROL GROUP

LISREL ESTIMATES (MAXIMUM LIKELIHOOD)

LAMBDA-Y

	ETA 1	ETA 2	ETA 3
VAR 1	1.00	- -	- -
VAR 2	1.00	0.70	- -
		(0.07)	
		9.56	
VAR 3	1.00	0.98	- -
		(0.08)	
		12.03	
VAR 4	1.00	1.00	- -
VAR 5	- -	- -	1.00
VAR 6	- -	- -	0.96
			(0.02)
			43.40

Although the results for the control group share some similarities with those observed in the experimental group (i.e., the loadings of the covariates on the Figural relations construct are quite similar to one another), there are some differences. Despite the fact that there is also growth in mean performance for the control group (i.e., the Shape factor loadings are significant), looking closely at the confidence intervals for THU_IND3 (0.82; 1.14), reveals that performance at THU_IND3 is the same as that at final assessment (i.e., THU_IND4). This is an aspect in which the two groups differ. Although in the experimental group there is a drop at the last assessment occasion, in the control group there is an indication of steady mean performance at the last two assessment occasions.

COVARIANCE MATRIX OF ETA

	LEVEL	SHAPE	FIG_REL1
LEVEL	293.33		
SHAPE	38.71	27.51	
FIGREL1	161.28	49.40	153.54

PSI

	LEVEL	SHAPE	FIG_REL1
LEVEL	293.33		
	(49.23)		
	5.96		

```
SHAPE              38.71        27.51
                  (17.13)      (12.61)
                    2.26         2.18

FIG_REL1          161.28        49.40       153.54
                  (31.00)      (13.53)      (27.76)
                    5.20         3.65         5.53
```

Unlike the experimental group, in the control group there is a significant (linear) relationship between Figural relations as a correlate of change along the Induction dimension. The covariance between the shape factor and initial Figural relations is significant: parameter estimate = 49.40, standard error = 13.53, and t value = 3.65. This group difference may be explained again by the fact that there was no training in the control group, whereas in the experimental group its effect may have capitalized on the particular modules of Induction-related tutor-guided instructions.

```
    THETA-EPS

          THU_IND1    THU_IND2    THU_IND3    THU_IND4      FR11        FR12
          --------    --------    --------    --------     -------     -------
             30.72       30.72       30.72       30.72      104.14       22.36
            (3.31)      (3.31)      (3.31)      (3.31)     (18.69)      (9.66)
             9.27        9.27        9.27        9.27        5.57        2.32

    SQUARED MULTIPLE CORRELATIONS FOR Y - VARIABLES

          THU_IND1    THU_IND2    THU_IND3    THU_IND4      FR11        FR12
          --------    --------    --------    --------     -------     -------
              .91         .92         .93         .93        .60         .86

    ALPHA

           LEVEL       SHAPE     FIG_REL1
           -----       -----     --------
           38.56        9.40        52.02
          (1.94)      (1.00)       (1.72)
          19.88        9.37        30.24
```

As in the experimental group, overall mean performance improvement is significant in the control group. This is reflected in the mean of the Shape factor: estimate = 9.40, standard error = 1.00, and a significant t value = 9.37. This is perhaps a reflection of the practice effects at work in both groups, due to the repeated assessment with the same instruments. Nevertheless, it is interesting to compare the mean overall growth across groups, to find out if there is significant training gain in the experimental group over and above these practice effects. As shown in the next subsection, this comparison can be accomplished by examining restricted versions of the model. Another interesting comparison involves the Shape mean across the control and experimental groups. Because the two groups were randomly formed at pretest, their means were comparable then. Thus, by looking at the shape mean group differences (control = 9.40 and experimental = 15.23) one can examine if the overall change was higher in the experimental group. As it turns out, this is actually the case in this study; the 95% CI for the Shape mean in the control group is (7.40; 11.40) and completely to the left of that in the experimental group, (13.71; 16.76).

Studying Plausibility of Group Restrictions

As discussed in chapter 1, in general one can make more precise statements about parameter comparability across and within groups with more restrictive versions of any proposed model (as long as the model is found tenable as a means of data description).This is due to the fact that parameter estimates in a more restrictive model are generally associated with smaller standard errors and hence shorter confidence intervals (i.e., lead to more accurate statements). This is part of the reason for looking for more parsimonious models that in addition have higher degrees of freedom and hence are associated with more dimensions along which they can be disconfirmed.

In the two-group analyses of the cognitive intervention data, one can further gain in model parsimony by observing that the groups were randomly formed in this empirical study after the pretest was conducted. Thus, no differences are expected across groups in any of the variable characteristics at pretest. That is, one would expect that, due to this reason, if group equality restrictions were imposed on variances, covariances, means, and covariate factor loadings at pretest, the fit of the model would not deteriorate markedly. This is precisely what the nested two-group model fitted next will do. In this constrained model all details pertaining to the pretests of Induction and Figural relations will be fixed for equality across groups—the factor loadings of the two covariate indicators, the variances and covariance of their construct and the level factor, as well as their two means. This is accomplished by adding the following LISREL command lines (one per constrained parameter) immediately before the last line of the whole input.

```
EQ PS(1, 1, 1) PS(1, 1)
EQ PS(1, 3, 3) PS(3, 3)
EQ PS(1, 3, 1) PS(3, 1)
EQ LY(1, 6, 3) LY(6, 3)
EQ AL(1, 1) AL(1)
EQ AL(1, 3) AL(3)
```

In each of these new command lines, a parameter from Group 1 (the experimental group) is fixed to equal the same parameter in Group 2 (the control group). In LISREL notation, this is accomplished by having three indices, where the first one always refers to the group number and the other two refer to the parameter location. Thus, PS(1, 1, 1) refers to the Group 1 element 1,1 of the PSI matrix. As such, the equality constraint EQ PS(1, 1, 1) PS(1, 1) sets the parameter of the current (control) group equal to that of the experimental group (i.e., Group 1). Using this notation, the first two command lines fix for group equality the variances of the Level and Figural relations factors (the first and third factors in the input), the third line fixes their covariance, and the fourth line fixes the factor loading of the second indicator of Figural relations. Finally, the means of the first and third factors (i.e., the Level and Figural relations factors) are set for group identity. (Note the reference by only two indices because the pertinent matrix ALpha is actually only a row and hence one needs to refer just to the position of the parameter of interest within it.) Each of the six parameters fixed in this way pertains to a characteristic of a variable assessed at pretest. Because the groups were randomly formed after the pretest, it seems reasonable to assume that these characteristics are identical across groups; this is being examined by testing these six restrictions.

This constrained model is associated with a chi-square value $T = 33.05$ with $df = 30, p = .32$, and an RMSEA $= .020$ with a 90% confidence interval $(0; .054)$. All these indices suggest an acceptable model fit. For future reference, call this model M_1 and the one fitted immediately before it M_0. M_1 is nested in M_0 because it can be obtained from the first model by imposing the proposed six parameter restrictions. Thus, M_1 has six more degrees of freedom because it has six fewer parameters than M_0.

The difference in chi-squares between the two nested models M_1 and M_0 is $\Delta T = 33.05 - 27.33 = 5.72$ with $\Delta df = 6$. Because the cut-off of the pertinent chi-square distribution with 6 degrees of freedom is 12.59 (at significance level .05, and obviously higher for any smaller level), this difference is determined to be nonsignificant. This implies that the imposed constraints are plausible, and can be retained in the ensuing analyses.

Now also examine evidence in this data favoring group invariance of the relationships among the three latent variables. To obtain the input for this model nested in M_1 (and thus in M_0 as well), one only needs to change the keyword PS=PS to PS=IN (for group INvariant) in the second-group model command line of the input. This constraint yields an increase in the chi-square value up to $T = 44.54$ with $df = 33$. Thus, the resulting difference in chi-squares is $\Delta T = 44.54 - 33.05 = 11.49$ with $\Delta df = 3$, and is significant (even at a conservative significance level .01, which one may want to adopt for protection in this multiple testing procedure with an unknown overall significance level). Hence, one can conclude that it is not likely that the same latent correlation matrix applies for both groups in the population. That is, the correlation structure among the Level, Shape, and Figural relation factors is not group invariant. Interestingly, even adding the constraint of identical latent means (instead of the incorrect group restrictions on all latent covariances and variances), accomplished by changing the keyword AL=PS to AL=IN in the control group, yields the same outcome. The chi-square value once again goes up to $T = 62.02$ with $df = 31$, and the chi-square difference test yields $\Delta T = 62.02 - 33.05 = 28.97$ with $\Delta df = 1$, which is also significant. One can therefore conclude that the latent mean of the Shape factor is not the same across groups (because this is the only latent mean not already set as group invariant in model M_1). Finally, restraining instead the factor loadings of the Shape factor across the groups (by changing the keyword LY=PS to LY=IN in the control group) yields $\Delta T = 46.71 - 33.05 = 13.66$ with $\Delta df = 2$, which is similarly significant, indicating that these loadings are not group invariant. Thus, the two groups differ also in the proportion of overall ability change that occurred from the pretest to first and second posttest assessment occasions.

Here the output of model M_1, the most restrictive and tenable of the proposed models, is presented (again, comments are inserted at appropriate places and repetitive material presented earlier in this section is omitted).

```
EXPERIMENTAL GROUP

PARAMETER SPECIFICATIONS

LAMBDA-Y
```

	LEVEL	SHAPE	FIG_REL1
THU_IND1	0	0	0
THU_IND2	0	1	0
THU_IND3	0	2	0
THU_IND4	0	0	0
FR11	0	0	0
FR12	0	0	3

```
     PSI
```

	LEVEL	SHAPE	FIG_REL1
LEVEL	4		
SHAPE	5	6	
FIG_REL1	7	8	9

```
     THETA-EPS
```

THU_IND1	THU_IND2	THU_IND3	THU_IND4	FR11	FR12
10	10	10	10	11	12

ALPHA

	LEVEL	SHAPE	FIG_REL1
	13	14	15

CONTROL GROUP

PARAMETER SPECIFICATIONS

LAMBDA-Y

	LEVEL	SHAPE	FIG_REL1
THU_IND1	0	0	0
THU_IND2	0	16	0
THU_IND3	0	17	0
THU_IND4	0	0	0
FR11	0	0	0
FR12	0	0	3

PSI

	LEVEL	SHAPE	FIG_REL1
LEVEL	4		
SHAPE	18	19	
FIG_REL	17	20	9

THETA-EPS

THU_IND1	THU_IND2	THU_IND3	THU_IND4	FR11	FR12
21	21	21	21	22	23

ALPHA

	LEVEL	SHAPE	FIG_REL1
	13	24	15

Note the identical numbers attached to the parameters constrained for equality across the two groups. Thus, model M_1 has, altogether, 24 parameters in both groups, six of which are set to be identical in the experimental and control groups.

EXPERIMENTAL GROUP

LISREL ESTIMATES (MAXIMUM LIKELIHOOD)

LAMBDA-Y

	LEVEL	SHAPE	FIG_REL1
THU_IND1	1.00	- -	- -

THU_IND2	1.00	1.03	- -
		(0.03)	
		31.84	
THU_IND3	1.00	1.14	- -
		(0.03)	
		33.15	
THU_IND4	1.00	1.00	- -
FR11	- -	- -	1.00
FR12	- -	- -	0.97
			(0.01)
			75.07

These parameters and standard errors are essentially the same as those discussed before for model M_0 and hence lead to the same interpretations. Only the standard error of the second Figural relations indicator loading is slightly smaller, but this is a finding of marginal relevance in these analyses, which are primarily concerned with studying change over time and the relationships among latent variables.

COVARIANCE MATRIX OF ETA

	LEVEL	SHAPE	FIG_REL1
LEVEL	287.41		
SHAPE	5.70	51.73	
FIG_REL1	175.72	16.45	151.08

PSI

	LEVEL	SHAPE	FIG_REL1
LEVEL	287.41		
	(28.16)		
	10.21		
SHAPE	5.70	51.73	
	(12.15)	(9.16)	
	0.47	5.65	
FIG_REL1	175.72	16.45	151.08
	(18.64)	(9.04)	(16.16)
	9.43	1.82	9.35

Note again that in the experimental group there does not seem to be a (linear) relationship between the Level and Shape factors—the estimate of their covariance is 5.70, with a standard error = 12.16 and a t value = 0.47, which is nonsignificant. That is, for experimental subjects the starting position on Induction is not predictive of the amount of change along that dimension. Similarly, there is a weak evidence of relatedness between the Shape and Figural relations factors—their covariance is estimated at 16.46, with a standard error = 9.04 and a t value = 1.82, which is not significant. Thus, in the experimental group there is a weak (linear) relationship between starting position on Figural relations and change along the Induction dimension. As mentioned before, this may, of course, have been an effect of the specific modules of the cognitive training. Nevertheless, in this model the statements are more precise due to the model's having more degrees of freedom than M_0. Specifically, the standard errors of all covariance estimates are on average markedly lower in M_0 than in M_1, which generally allows more accurate conclusions.

THETA-EPS

	THU_IND1	THU_IND2	THU_IND3	THU_IND4	FR11	FR12
	22.08	22.08	22.08	22.08	81.18	24.77
	(1.74)	(1.74)	(1.74)	(1.74)	(11.19)	(6.51)
	12.68	12.68	12.68	12.68	7.25	3.81

SQUARED MULTIPLE CORRELATIONS FOR Y - VARIABLES

	THU_IND1	THU_IND2	THU_IND3	THU_IND4	FR11	FR12
	0.93	0.94	0.94	0.94	0.65	0.85

* CONTROL GROUP

LISREL ESTIMATES (MAXIMUM LIKELIHOOD)

LAMBDA-Y

	LEVEL	SHAPE	FIG_REL1
THU_IND1	1.00	- -	- -
THU_IND2	1.00	0.69	- -
		(0.08)	
		9.02	
THU_IND3	1.00	0.98	- -
		(0.09)	
		11.36	
THU_IND4	1.00	1.00	- -
FR11	- -	- -	1.00
FR12	- -	- -	0.97
			(0.01)
			75.07

As can be seen in the control group output, the findings are essentially the same as the ones with model M_0 discussed previously.

COVARIANCE MATRIX OF ETA

	LEVEL	SHAPE	FIG_REL1
LEVEL	287.41		
SHAPE	48.14	23.82	
FIG_REL1	175.72	47.44	151.08

PSI

	LEVEL	SHAPE	FIG_REL1
LEVEL	287.41		
	(28.16)		
	10.21		
SHAPE	48.14	23.82	
	(15.17)	(12.05)	
	3.17	1.98	
FIG_REL1	175.72	47.44	151.08
	(18.64)	(11.71)	(16.16)
	9.43	4.05	9.35

THETA-EPS

THU_IND1	THU_IND2	THU_IND3	THU_IND4	FR11	FR12
31.46	31.46	31.46	31.46	105.00	29.34
(3.37)	(3.37)	(3.37)	(3.37)	(18.61)	(9.17)
9.34	9.34	9.34	9.34	5.64	3.20

SQUARED MULTIPLE CORRELATIONS FOR Y - VARIABLES

THU_IND1	THU_IND2	THU_IND3	THU_IND4	FR11	FR12
0.90	0.92	0.93	0.93	0.59	0.83

ALPHA

LEVEL	SHAPE	FIG_REL1
37.87	9.02	50.52
(1.13)	(0.91)	(0.98)
33.60	9.90	51.50

Unlike the experimental group, in the control group one again finds evidence of a significant relationship between the Level and Shape factors—their covariance is estimated at 48.14, with a standard error = 15.17 and a *t* value = 3.17, which is significant. The correlation of the two factors using the values of the variance and covariance estimates in the PSI matrix is computed to be equal to .58, indicating a moderately strong (linear) relationship. This suggests that in the control group (in which subjects did not receive any information about how well they were doing) individuals starting high on Induction tended to be among those exhibiting the most improvement along this dimension. Furthermore, the relationship between initial Figural relations and change along the Induction dimension is also significant in the control group (i.e., the covariance is estimated at 47.45, with a standard error = 11.72 and t value = 4.05, which is significant). This leads to a correlation that is equal to .79, indicating a strong relationship. Thus, individuals in the control group that start high on Figural relations tend to be among those improving most along the Induction dimension.

Finally, the fit indices of model M_1 are all satisfactory and hence, as illustrated previously, one can proceed with a discussion of parameter interpretation.

```
                    GOODNESS OF FIT STATISTICS

    CHI-SQUARE WITH 30 DEGREES OF FREEDOM = 33.05 (P = 0.32)
                CONTRIBUTION TO CHI-SQUARE = 8.63
           PERCENTAGE CONTRIBUTION TO CHI-SQUARE = 26.13
          ESTIMATED NON-CENTRALITY PARAMETER (NCP) = 3.05
       90 PERCENT CONFIDENCE INTERVAL FOR NCP = (0.0 ; 21.29)

                MINIMUM FIT FUNCTION VALUE = 0.13
       POPULATION DISCREPANCY FUNCTION VALUE (F0) = 0.012
        90 PERCENT CONFIDENCE INTERVAL FOR F0 = (0.0 ; 0.087)
     ROOT MEAN SQUARE ERROR OF APPROXIMATION (RMSEA) = 0.020
     90 PERCENT CONFIDENCE INTERVAL FOR RMSEA = (0.0 ; 0.054)
        P-VALUE FOR TEST OF CLOSE FIT (RMSEA < 0.05) = 0.92

        EXPECTED CROSS-VALIDATION INDEX (ECVI) = 0.33
      90 PERCENT CONFIDENCE INTERVAL FOR ECVI = (0.18 ; 0.27)
                ECVI FOR SATURATED MODEL = 0.17
               ECVI FOR INDEPENDENCE MODEL = 8.09

 CHI-SQUARE FOR INDEPENDENCE MODEL WITH 30 DEGREES OF FREEDOM = 1978.73
                 INDEPENDENCE AIC = 2002.73
                      MODEL AIC = 81.05
                    SATURATED AIC = 84.00
                 INDEPENDENCE CAIC = 2056.89
                     MODEL CAIC = 189.37
                   SATURATED CAIC = 273.56

          ROOT MEAN SQUARE RESIDUAL (RMR) = 11.16
                  STANDARDIZED RMR = 0.036
            GOODNESS OF FIT INDEX (GFI) = 0.97
       PARSIMONY GOODNESS OF FIT INDEX (PGFI) = 1.39

              NORMED FIT INDEX (NFI) = 0.98
            NON-NORMED FIT INDEX (NNFI) = 1.00
         PARSIMONY NORMED FIT INDEX (PNFI) = 0.98
            COMPARATIVE FIT INDEX (CFI) = 1.00
            INCREMENTAL FIT INDEX (IFI) = 1.00
             RELATIVE FIT INDEX (RFI) = 0.98
                  CRITICAL N (CN) = 379.82
```

References

Akaike, H. (1987). Factor analysis and AIC. *Psychometrika, 52*, 317–332.

Allen, M. J., & Yen, W. M. (1979). *Introduction to measurement theory.* Belmont, CA: Wadsworth.

Arbuckle, J. L. (1995). *Amos user's guide.* Chicago: Smallwaters.

Babbie, E. (1992). *The practice of social research.* Belmont, CA: Wadsworth.

Baltes, P. B., Dittmann-Kohli, F., & Kliegl, R. (1986). Reserve capacity of the elderly in aging sensitive tests of fluid intelligence: Replication and extension. *Psychology and Aging, 1*, 172–177.

Bentler, P. M. (1990). Comparative fit indexes in structural equation models. *Psychological Bulletin, 107*, 238–246.

Bentler, P. M. (1995). *EQS structural equations program manual.* Encino, CA: Multivariate Software.

Bentler, P. M. (2000). *EQS 6 Structural Equation Program Manual.* Encino, CA: Multivariate Software.

Bentler, P. M., & Bonnett, D. (1980). Significance tests and goodness of fit in the analysis of covariance structures. *Psychological Bulletin, 88*, 588–606.

Bollen, K. A. (1989). *Structural equations with latent variables.* New York: Wiley.

Breckler, S. J. (1990). Application of covariance structure analysis modeling in psychology: A cause of concern? *Psychological Bulletin, 107*, 260–273.

Browne, M. W., & Cudeck, R. (1993). Alternative ways of assessing model fit. In K. A. Bollen & J. S. Long (Eds.), *Testing structural equation models* (pp. 136–162). Newbury Park, CA: Sage.

Browne, M. W., & Mels, G. (1994). *RAMONA PC user's guide.* Columbus, OH: Department of Psychology, Ohio State University.

Byrne, B. M. (1998). *Structural equation modeling with LISREL, PRELIS, and SIMPLIS: Basic concepts, applications, and programming.* Mahwah, NJ: Lawrence Erlbaum Associates.

Crocker, L., & Algina, J. (1986). *Introduction to classical and modern test theory.* New York: Harcourt Brace Jovanovich.

Cudeck, R. (1989). Analyzing correlation matrices using covariance structure models. *Psychological Bulletin, 105*, 317–327.

Drezner, Z., Marcoulides, G. A., & Salhi, S. (1999). Tabu search model selection in multiple regression models. *Communications in Statistics, 28*(2), 349–367.

Duncan, T. E., Duncan, S. C., Strycker, L. A., Li, F., & Alpert, A. (1999). *An introduction to latent variable growth curve modeling.* Mahwah, NJ: Lawrence Erlbaum Associates.

Finn, J. D. (1974). *A general model for multivariate analysis.* New York: Holt, Reinhart & Winston.

Hays, W. L. (1994). *Statistics.* Fort Worth, TX: Holt, Rinehart & Winston.

Heck, R. H., & Johnsrud, L. K. (1994). Workplace stratification in higher education administration: Proposing and testing a structural model. *Structural Equation Modeling, 1*(1), 82–97.

Horn, J. L. (1982). The aging of human abilities. In B. B. Wolman (Ed.), *Handbook of developmental psychology* (pp. 847–870). New York: McGraw-Hill.

Hu, L.-T., & Bentler, P. M. (1999). Cutoff criteria for fit indices in covariance structure analysis: Conventional criteria versus new alternatives. *Structural Equation Modeling, 6*(1), 1–55.

Hu, L.-T., Bentler, P. M., & Kano, Y. (1992). Can test statistics in covariance structure analysis be trusted? *Psychological Bulletin, 112,* 351–362.

Huitema, B. J. (1980). *Analysis of covariance and its alternatives.* New York: Wiley.

Jöreskog, K. G. (1971). Statistical analysis of sets of congeneric tests. *Psychometrika, 36,* 109–133.

Jöreskog, K. G., & Sörbom, D. (1990). Model search with TETRAD II and LISREL. *Sociological Methods and Research, 19*(1), 93–106.

Jöreskog, K. G., & Sörbom, D. (1993a). *LISREL8: The SIMPLIS command language.* Chicago: Scientific Software.

Jöreskog, K. G., & Sörbom, D. (1993b). *LISREL8: User's reference guide.* Chicago: Scientific Software.

Jöreskog, K. G., & Sörbom, D. (1993c). *PRELIS2: A preprocessor for LISREL.* Chicago: Scientific Software.

Jöreskog, K. G., & Sörbom, D. (1999). *LISREL8.30: User's reference guide.* Chicago: Scientific Software.

Long, J. S. (1983). *Covariance structure models: An introduction to LISREL.* Beverly Hills, CA: Sage.

Marcoulides, G. A. (1989). Structural equation modeling for scientific research. *Journal of Business and Society, 2*(2), 130–138.

Marcoulides, G. A. (Ed.). (1998). *Modern methods for business research.* Mahwah, NJ: Lawrence Erlbaum Associates.

Marcoulides, G. A., & Drezner, Z. (in press). Specification searches in structural equation modeling with a genetic algorithm. In G. A. Marcoulides & R. E. Schumacker (Eds.). *Advanced structural equation modeling: New developments and techniques.* Mahwah, NJ: Lawrence Erlbaum Associates.

Marcoulides, G. A., & Drezner, Z. (1999). Using simulated annealing for model selection in multiple regression analysis. *Multiple Linear Regression Viewpoints, 25*(2), 1–4.

Marcoulides, G. A., Drezner, Z., & Schumacker, R. E. (1998). Model specification searches in structural equation modeling using Tabu search. *Structural Equation Modeling, 5*(4), 365–376.

Marcoulides, G. A., & Hershberger, S. L. (1997). *Multivariate statistical methods.* Mahwah, NJ: Lawrence Erlbaum Associates.

Marcoulides, G. A., & Schumacker, R. E. (1996). *Advance structural equation modeling: Issues and techniques.* Mahwah, NJ: Lawrence Erlbaum Associates.

Marsh, H. W., Balla, J. R., & Hau, K.-T. (1996). An evaluation of incremental fit indices: A clarification of mathematical and empirical properties. In G. A. Marcoulides & R. E.

Schumacker (Eds.), *Advanced structural equation modeling: Issues and techniques* (pp. 325–353). Mahwah, NJ: Lawrence Erlbaum Associates.

McArdle, J. J. (1988). Dynamic but structural equation modeling of repeated measures data. In J. R. Nesselroade & R. B. Cattell (Eds.), *Handbook of multivariate experimental psychology* (2nd ed., pp. 561–614). New York: Plenum.

McArdle, J. J. (1998). Modeling longitudinal data by latent growth curve methods. In G. A. Marcoulides (Ed.), *Modern methods for business research* (pp. 359–406). Mahwah, NJ: Lawrence Erlbaum Associates.

McArdle, J. J., & Anderson, E. (1990). Latent variable growth models for research on aging. In J. E. Birren & K. W. Schaie (Eds.), *Handbook of the psychology of aging* (3rd ed., pp. 21–44). New York: Academic Press.

McArdle, J. J., & Epstein, D. (1987). Latent growth curves within developmental structural equation models. *Child Development, 58,* 110–133.

Meredith, W., & Tisak, J. (1990). Latent curve analysis. *Psychometrika, 55,* 107–122.

Muthén, B., & Muthén, L. (1998). *MPLUS user's guide.* Los Angeles: Muthén & Muthén.

Pedhauzer, E. J., & Schmelkin, L. (1991). *Measurement, design, and analysis: An integrated approach.* Hillsdale, NJ: Lawrence Erlbaum Associates.

Popper, K. (1962). *On the structure of scientific revolutions.* Chicago: Chicago University Press.

Raykov, T. (1994). Studying correlates and predictors of longitudinal change using structural modeling. *Applied Psychological Measurement, 17,* 63–77.

Raykov, T. (1995). Multivariate structural modeling of plasticity in fluid intelligence of aged adults. *Multivariate Behavioral Research, 30,* 255–288.

Raykov, T. (in press). On sensitivity of structural equation modeling to latent relationship misspecification. *Structural Equation Modeling.*

Raykov, T., & Marcoulides, G. A. (1999). On desirability of parsimony in structural equation model selection. *Structural Equation Modeling, 6,* 292–300.

Raykov, T., & Marcoulides, G. A. (in press). Can there be infinitely many models equivalent to a given covariance structure model? *Structural Equation Modeling.*

Raykov, T., & Penev, S. (1999). On structural equation model equivalence. *Multivariate Behavioral Research, 34,* 199–244.

Raykov, T., & Widaman, K. F. (1995). Issues in applied structural equation modeling research. *Structural Equation Modeling, 2,* 289–318.

Rigdon, E. E. (1998). Structural equation modeling. In G. A. Marcoulides (Ed.), *Modern methods for business research* (pp. 251–294). Mahwah, NJ: Lawrence Erlbaum Associates.

SAS Institute. (1989). *SAS PROC CALIS user's guide.* Cary, NC: Author.

Scheines, R., Spirtes, P., & Glymour, C. (in press). Model specification searches. In G. A. Marcoulides & R. E. Schumacker (Eds.), *Advanced structural equation modeling: New developments and techniques.* Mahwah, NJ: Lawrence Erlbaum Associates.

Scheines, R., Spirtes, P., Glymour, G., Meek, C., & Richardson, T. (1998). The TETRAD project: Constraint based aids to causal model specification. *Multivariate Behavioral Research, 33*(1), 65–117.

Schumacker, R. E., & Lomax, R. G. (1996). *A beginner's guide to structural equation modeling.* Mahwah, NJ: Lawrence Erlbaum Associates.

Schumacker, R. E., & Marcoulides, G. A. (Eds.). (1998). *Interaction and nonlinear effects in structural equation modeling.* Mahwah, NJ: Lawrence Erlbaum Associates.

Spirtes, P., Scheines, R., & Glymour, C. (1990). Simulation studies of the reliability of computer aided specification using the TETRAD II, EQS, and LISREL programs. *Sociological Methods and Research, 19*(1), 3–66.

Statistica (1998). *User's guide.* Tucson, AZ: Author.

Steiger, J. H. (1998). A note on multiple sample extensions of the RMSEA fit index. *Structural Equation Modeling, 5*, 411–419.

Steiger, J. H. & Lind, J. C. (1980, June). *Statistically based tests for the number of common factors.* Paper presented at the Psychometric Society Annual Meeting, Iowa City, IA.

Suen, H. K. (1990). *Principles of test theories.* Hillsdale, NJ: Lawrence Erlbaum Associates.

Tabachnick, B. G., & Fidell, L. S. (1999). *Using multivariate statistics.* New York: HarperCollins.

Thurstone, L. L. (1935). *The vectors of the mind.* Chicago: University of Chicago Press.

Willett, J. B., & Sayer, A. G. (1996). Cross-domain analyses of change over time: Combining growth modeling and covariance structure analysis. In G. A. Marcoulides & R. E. Schumacker (Eds.), *Advanced structural equation modeling: Issues and techniques* (pp. 125–157). Mahwah, NJ: Lawrence Erlbaum Associates.

Wolfle, L. M. (1999). Sewall Wright on the method of path coefficients: An annotated bibliography. *Structural Equation Modeling, 6*(3), 280–291.

Wright, S. (1920). The relative importance of heredity and environment in determining the piebald pattern of guinea-pigs. *Proceedings of the National Academy of Sciences, 6*, 320–332.

Wright, S. (1921). Correlation and causation. *Journal of Agricultural Research, 20*, 557–585.

Wright, S. (1934). The method of path coefficients. *Annals of Mathematical Statistics, 5*, 161–215.

Author Index

A

Akaike, H., 41, 197
Algina, J., 34, 197
Allen, M. J., 34,197
Alpert, A., 148, 149, 198
Anderson, E., 165, 166, 199
Arbuckle, J. L., 2, 197

B

Babbie, E., 197, 10
Balla, J. R., 2, 40, 198
Baltes, P. B., 152, 179, 197
Bentler, P. M., 2, 3, 15, 26, 27, 33, 38,
    39, 41, 46, 48, 154, 156, 197
Bollen, K. A., 10, 27, 33, 39, 197
Bonnett, D., 38, 197
Breckler, S.J., 37, 197
Browne, M. W., 2, 40, 41, 197
Byrne, B. M., 23, 197

C

Crocker, L., 34, 197
Cudeck, R., 40, 41, 197

D

Dittmann-Kohli, F., 152, 179, 197

D

Drezner, Z., 44, 45, 90, 91, 198
Duncan, S. C., 148, 149, 198
Duncan, T. E., 148, 149, 198

E

Epstein, D., 148, 199

F

Fidell, L. S., 27, 147, 200
Finn, J. D., 64, 198

G

Glymour, C., 44, 90, 199

H

Hau, K.-T., 2, 40, 198
Hays, W. L., 20, 150, 151, 198
Heck, R. H., 5, 198
Hershberger, S. L., 3, 5, 24, 27, 33, 39,
    111, 147, 198
Horn, J. L., 122, 198
Hu, L.-T., 27, 38, 39, 198
Huitema, B. J., 148, 198

Subject Index

A

ADF (asymptotically distribution free), *see* Estimators

Adjusted goodness of fit index, *see* Evaluation of fit

AIC, *see* Evaluation of fit

Akaike's criterion, *see* Evaluation of fit

Alpha matrix, 154–156

AMOS, *see* Computer programs

ANCOVA, 147–148

ANOVA, 147

Augmented matrix, 150–157

B

Beta matrix, 57

C

CFI, *see* Evaluation of fit

Chi-square, *see* Evaluation of fit

Computer programs, 1, 46–62

    AMOS, 1

    EQS, 1, 46–62

    LISREL, 1, 46–62

    Mplus, 1

    PRELIS, 1

    SAS (PROC CALIS), 1

    STATISTICA (SEPATH), 1

    RAMONA, 1

Confidence intervals, 29–30, 171–172

Confirmatory factor analysis, 4, 95–120

Consistent AIC, *see* Evaluation of fit

Constrained parameters, *see* Model parameters

Construct validation, 6

D

Degrees of freedom, 32

Dependent variables, 11–12

Determininant of matrix, 71

Direct effects, 7

Dummy variable, 156–157

E

Endogenous variables, 11

Equivalent models, 35

EQS, *see* Computer programs

Estimators, 26–30

    asymptotically distribution free, 26–27

    generalized least squares, 26

    least squares, 26–27

    maximum likelihood, 26–29

    weighted least squares, 26–27